Frontiers in Applied Dynamical Systems: Reviews and Tutorials

Volume 7

More information about this series at http://www.springer.com/series/13763

Frontiers in Applied Dynamical Systems: Reviews and Tutorials

Frontiers in Applied Dynamical Systems covers emerging topics and significant developments in the field of Dynamical Systems. It is an annual collection of invited review articles by leading researchers in dynamical systems and related areas. Contributions in this series should be seen as a portal for a broad audience of researchers in dynamical systems at all levels and can serve as advanced teaching aids. Each contribution provides an informal outline of a specific area, an interesting application, a recent technique, or a "how-to" for analytic methods and for computational algorithms, and a list of key references. All articles will be refereed.

Mike R. Jeffrey

Modeling with Nonsmooth Dynamics

Springer

Mike R. Jeffrey
Department of Engineering Mathematics
University of Bristol
Bristol, Avon, UK

ISSN 2364-4532 ISSN 2364-4931 (electronic)
Frontiers in Applied Dynamical Systems: Reviews and Tutorials
ISBN 978-3-030-35986-7 ISBN 978-3-030-35987-4 (eBook)
https://doi.org/10.1007/978-3-030-35987-4

Mathematics Subject Classification: 00A69, 00A71, 00A72, 34-02, 34A26, 34A38, 34A60, 34C60, 34D15, 34E05, 37-02, 37N99

This Springer imprint is published by the registered company Springer Nature Switzerland AG.
The registered company address is: Gewerbestrasse 11, 6330 Cham, Switzerland

Preface

As mathematics is applied to model ever new problems in engineering and the life sciences, increasing use is being made of systems that switch between different sets of equations on distinct domains. To find their dynamics requires the discontinuity between domains to be resolved or 'regularized' in some way, and there exist a range of methods to do so. Some preserve the ideal character of the discontinuity as a piecewise-smooth system (giving, e.g. 'impact' or 'switching' dynamics), while others blur the discontinuity by smoothing it out, or introducing overshoots due to deterministic or stochastic delays.

Despite exciting new applications and major theoretical advances, it remains unclear how widely applicable nonsmooth models are, or in what sense they approximate discontinuities in real-world systems. It is even unclear how to correctly simulate or solve nonsmooth systems, or how robust such solutions are to perturbation. To move closer towards these goals, here we survey one of the main approaches to modelling nonsmooth dynamics, and look at how loosening some of its rigourous but idealized framework allows us to probe its modelling assumptions. We also draw together a range of phenomena that characterize the sensitivity and robustness of nonsmooth dynamical models.

Bristol, UK Mike R. Jeffrey

Contents

Chapter 1
Mathematics for a Nonsmooth World

A system may be said to exhibit *nonsmooth dynamics* if the laws that govern its behaviour change markedly at certain thresholds. Those changes might represent decisions, physical switches, boundaries of solid objects, or changes in modes of contact. Here we look at the current understanding of nonsmooth dynamics as a means of approximation, and how it allows us to model novel phenomena beyond the scope of smooth dynamical systems.

Recent years have seen a huge growth in both the theory and applications of nonsmooth dynamics. The decision of an investor to change their trading tactics, of a predator to change their prey, of a neuron to fire or a cell to divide, are all forms of nonsmooth dynamics. When you apply the brakes in your car or collide with a wall, or when you abandon your regular commute in favour of your travel apps' suggested alternative, these all induce nonsmooth dynamics of some kind, and have led the discipline far beyond its original home in contact mechanics and electronic control. We review some example applications in Chap. 2.

Nonsmooth dynamical systems are often invoked with a notable degree of unease, a lingering discomfort of the conceptual difficulties of marrying together differentiable and discontinuous change. By very definition, the far-reaching theorems of differentiable dynamical systems do not apply to nonsmooth events. All we know about stability, attraction, bifurcations, and chaos have to be re-written for these nonsmooth, or more precisely *piecewise-smooth*, dynamical systems.

The last 40 years have seen leaps forward in tackling these challenges. Propelled forward largely by the separate works of A. F. Filippov, M. A. Teixeira, and M.I. Feigin, we are discovering how to rigorously extend the ideas of stability—which are so important to differentiable dynamical systems—to thresholds where differentiability fails. The translation of Filippov's book [51] stands as the most extensive development of singularity theory and qualitative dynamics for 'differential equations with discontinuous right-hand sides', formalizing the important notion of *sliding* along a discontinuity, itself the culmination of decades of work in the Russian literature such as [3, 4, 7, 50, 108], and sparking off V. I. Utkin's widely known "*variable structure systems*" for electronic control, see, e.g., [127, 146, 147]. Teixeira's body

M. R. Jeffrey, *Modeling with Nonsmooth Dynamics*,
Frontiers in Applied Dynamical Systems: Reviews and Tutorials 7,
https://doi.org/10.1007/978-3-030-35987-4_1

of work, [138, 139] to [140], has led the way into higher dimensions and exploring the connection to singular perturbed systems. While these contributed primarily to local dynamics, Feigin introduced the notion of *C-bifurcations* in [44, 48] to describe global bifurcation of orbits grazing a discontinuity. These advances directly led to the recent explosion in theories concerning regularization of discontinuities, equivalence classes of similar systems and bifurcations between them, and the existence of limit cycles; we discuss these to some extent in Chap. 2.

Despite the growth in applications and advances in theory, nagging and important problems about the uniqueness of nonsmooth models refuse to go away. The solutions of equations like

$$\text{(a)} \quad \tfrac{dx}{dt} = \sqrt{|x|} \quad \text{or} \quad \text{(b)} \quad \tfrac{dx}{dt} = 2 + \tfrac{x}{|x|} \tag{1.1}$$

evolve into and out of the point $x = 0$ in finite time, but do they pass through $x = 0$ or can they stop there? Non-differentiable situations like these are created easily by writing step functions, inverse functions, or IF statements in computational code, and such terms arise commonly in applications, but the behaviour they result in can vary hugely depending on how the resulting discontinuity is handled. It turns out that we can utilize such non-uniqueness to obtain useful solutions and increase the reach of our mathematical models. These are the less well explored issues of nonsmooth dynamics that this article attempts to address.

The often overlooked fact is that if discontinuous quantities appear in a differential equation, then its dynamical behaviour is necessarily non-unique. Standard texts on dynamical systems tend to highlight examples like (1.1) as warnings to avoid nonsmooth systems. Filippov and his contemporaries' work showed that solutions to discontinuous problems exist and can be studied just as meaningfully as differentiable systems, but the tendency remained to attempt to find conditions that banish non-uniqueness. A currently popular trend of smoothing out discontinuities (to 'regularize' them), particularly following [136], makes it clear that nonsmoothness is intimately tied to the concept of singular perturbations, yet regularizations are still largely studied with the aim of banishing non-uniqueness. As discussed in [81, 82], the link to singular perturbations actually points to non-uniqueness having a more important role, and [79, 81] proposed how this can and should be utilized to avoid misleading analysis, and to express the different behaviours made possible by discontinuous quantities. Starting with Sect. 1.1 we shall begin further teasing out the connection to singular perturbations and non-unique limits.

At stake is our entire understanding of what a discontinuity constitutes as an approximation. Are nonsmooth systems a legitimate modelling tool or an unmathematical fudge? Are they reliable, and what kinds of perturbation are they robust to? Our aim is to bring these open questions to the fore and to help begin formulating them in a useful way. It would be premature to attempt to answer these questions definitively, instead we introduce a framework in a largely informal manner, aiming at helping to move the discussion forward.

We will reveal how sensitive nonsmooth models are to real world non-idealities, particularly perturbations that delay or randomize the way a quantity switches in value. Such non-idealities result in complex solutions that are not expressible in closed form, but can still be brought within the powerful concepts set out by Filippov by loosening some of the key definitions—such as that of *sliding* along a discontinuity threshold—that have been instrumental to progress in the study of stability and bifurcations in nonsmooth models.

Before getting into these issues, let us spend the rest of Chap. 1 introducing some key ideas and applications of nonsmooth dynamics.

1.1 What Are Discontinuities Hiding?

Suppose that the variation of a quantity x in a system appears to exhibit behaviour that is discontinuous across the threshold $x = 0$, and we model this with a discontinuous function $F(x)$, writing $\dot{x} = F(x)$ where

$$F(x) = \begin{cases} a_0(x) & \text{if } \; x > 0 \, , \\ b_0(x) & \text{if } \; x < 0 \, . \end{cases} \tag{1.2}$$

We can assume for now that a_0 and b_0 are smooth functions defined at least on the closure of their domains in (1.2), i.e., up to and including the threshold $x = 0$. Given that our model provides no clear unambiguous value for $F(x)$ at $x = 0$, how should we solve the system at the discontinuity?

We may look to a letter written by George Gabriel Stokes to his fiancée in 1857 for inspiration:

> When the cat's away the mice may play. You are the cat and I am the mouse. I have been doing what I guess you won't let me do when we are married, sitting up till 3 o'clock in the morning fighting hard against a mathematical difficulty. Some years ago I attacked an integral of Airy's, and after a severe trial reduced it to a readily calculable form. But there was one difficulty about it which, though I tried till I almost made myself ill, I could not get over . . . the discontinuity of arbitrary constants. [95]

Thus it became known, much as it pained Stokes to discover, that the *asymptotic* expansion of integrals and differential equations could contain discontinuous quantities. Imagine we are trying to model a quantity whose rate of change obeys $\dot{x} = F$, and we know that F satisfies some relation $\mathrm{D}[t, x, F; \varepsilon] = 0$ in terms of some small parameter ε, where D represents an integral, or ordinary, partial, stochastic, or other differential operation on F with respect to x and t (we given an example below, and a few others are given in section 5 of [81]). In effect Stokes showed that in attempting to solve such a relation $\mathrm{D}[t, x, F; \varepsilon] = 0$, even if D was differentiable in each of its arguments, we could find that F has a different asymptotic approximation in different regimes, for example, that F behaves as

$$F(x) = \begin{cases} A(x, s(x); \varepsilon) & \text{for } \; x \gg +\varepsilon \, , \\ B(x, s(x); \varepsilon) & \text{for } \; x \ll -\varepsilon \, , \end{cases} \tag{1.3}$$

where $\varepsilon > 0$ is small and

$$A(x, s; \varepsilon) = \sum_{n=0}^{\infty} a_n(x; \varepsilon) s^n \quad \& \quad B(x, s; \varepsilon) = \sum_{n=0}^{\infty} b_n(x; \varepsilon) s^n \,, \tag{1.4}$$

in terms of smooth functions $a_n(x; \varepsilon)$ and $b_n(x; \varepsilon)$ which remain regular for $\varepsilon \to 0$, so that

$$a_n(x) = a_n(x; 0) \quad \& \quad b_n(x) = b_n(x; 0) \tag{1.5}$$

are regular functions. The expansion variable s is a function of x and some small parameter ε such that $s = O(\varepsilon/x)$, typically just $s = \varepsilon/x$, or a term like $s = \mathrm{e}^{-|x|/\varepsilon}$ that is exponentially small in ε.

The problem confronting Stokes was to solve a differential equation $\varepsilon F'' = xF$ (our "$\mathrm{D}[t, x, F; \varepsilon] = 0$") for small ε, which described the diffraction of light near a caustic, for example, a rainbow. A similar but more convenient example is

$$\varepsilon^2 F'' = -xF' \,, \tag{1.6}$$

where $F' \equiv \frac{\partial F}{\partial x}$, and we will take convenient boundary conditions $F = c$ and $\varepsilon F' = \sqrt{2/\pi}$ at $x = 0$. (Recall that in our problem of interest F we also have $\dot{x} = F$ alongside this). The solution to (1.6) in this case is an integral,

$$\begin{aligned} F(x) &= c + \mathrm{Erf}\left(\frac{x}{\sqrt{2}\varepsilon}\right) \\ &= \begin{Bmatrix} c+1 \text{ for } x \gg +\varepsilon \\ c-1 \text{ for } x \ll -\varepsilon \end{Bmatrix} - \sqrt{\frac{2}{\pi}}\, \mathrm{e}^{-x^2/2\varepsilon^2} \sum_{p=0}^{\infty} \frac{(-1)^p (2p-1)!!}{(x/\varepsilon)^{2p+1}} \,, \end{aligned} \tag{1.7}$$

where Erf is the standard error function [1], and the second line expands its asymptotic form for large $|x|/\varepsilon$ (which we derive in Appendix A). Note that letting $c = 3$, for example, ignoring the higher order term and then setting $\varepsilon = 0$, we may write this as $\dot{x} = 2 + \frac{x}{|x|}$, i.e., the system in our earlier example (1b).

Since the appearance of such *Stokes discontinuities* is well understood, the suggestion was made in [81] that we might consider a nonsmooth system like (1.2) to in fact be an attempt to model a system with discontinuous asymptotics given by (1.3). Unlike the ideal fields of fundamental physics, however, in the messy worlds of engineering, biology, climate, and so on, one rarely possesses such ideal models as the Airy integral that Stokes had in his possession from which to derive (1.3), and must instead be content with observations leading directly to empirical models with forms like (1.2).

We can, however, use this association with asymptotics to refine the model (1.2) without access to the unknown laws behind it. We shall see that we have not done badly with (1.2), as it constitutes the leading order behaviour of (1.4).

A function with a Stokes discontinuity can be expressed as a uniform asymptotic expansion valid for all x, for example, from (1.3) it is obvious that we can write

$$F(x) = \nu A(x, s; \varepsilon) + (1 - \nu)B(x, s; \varepsilon) \,, \tag{1.8}$$

in terms of a quantity ν that equals 1 for $x \gg \varepsilon$ and 0 for $x \ll \varepsilon$. In (1.7), for example, we may write $F(x) = c - 1 + 2\nu + O(\varepsilon/x)$ (see Appendix A for the exact expression in the form of (1.8)). The term ν must therefore be asymptotically a step function, which we may write here as

$$\nu = \begin{Bmatrix} 1 - s \text{ for } x \gg +\varepsilon \\ s \quad\;\; \text{for } x \ll -\varepsilon \end{Bmatrix} + O\left(s^2\right) , \tag{1.9}$$

without loss of generality. (Universal forms for the multiplier ν are known for certain classes of equations, see particularly [16]). Using (1.9) we can eliminate s from these expressions altogether, as $s \sim 1 - \nu$ in $x \gg +\varepsilon$ and $s \sim \nu$ in $x \ll -\varepsilon$. Substituting these into (1.4) and (1.8), we obtain a series that can be factorized as

$$F(x) = \nu a_0(x; \varepsilon) + (1 - \nu)b_0(x; \varepsilon) + \nu(1 - \nu)G(x; \nu, \varepsilon) \,. \tag{1.10}$$

The first two terms are the leading order approximations in the regions $x \gg +\varepsilon$ and $x \ll -\varepsilon$. The higher order terms, which are small for all $|x| \gg \varepsilon$, go into the term $\nu(\nu - 1)G$, where G is a series made up of the remaining functions a_n and b_n for $n \geq 1$. We provide finer details of this derivation in Appendix A.

Now, however, there is no explicit mention of s in (1.11), and any remaining dependence on the small parameter ε is regular by (1.5). That is, if we let $\varepsilon \to 0$ in (1.9), we obtain a series expansion for F in polynomials of ν,

$$F(x) = \nu a_0(x) + (1 - \nu)b_0(x) + \nu(1 - \nu)G(x; \nu) \,. \tag{1.11}$$

and ν by (1.9) becomes exactly a discontinuous step function. The leading order of the expansion is linear in ν, which produces (1.2), while the higher order terms are nonlinear in ν and are *zero almost everywhere*! That is, since ν equals 1 or 0 for $x > 0$ or $x < 0$, the term $\nu(\nu - 1)G$ vanishes for all $x \gtrless 0$. We call this a *hidden* term. We even obtain a formula for G (whose derivation we give in Appendix A), namely

$$G(x; \nu) = \sum_{n=0}^{\infty} \{a_{n+1}(x)(1 - \nu)^n + b_{n+1}(x)\nu^n\} \,, \tag{1.12}$$

(recalling (1.5)). The expansion (1.7), for example, becomes

$$F(x) = c - 1 + 2\nu + O(\nu(1 - \nu)) \,. \tag{1.13}$$

We shall obtain other examples of the expansion (1.11), where different hidden terms (1.12) represent different smooth systems which all give the same discontinuous modes (1.2) in the limit $\varepsilon \to 0$.

Thus in (1.11) we have a series expansion, about a discontinuity at $x = 0$, in polynomials of a discontinuous quantity ν, that approximates about the known values of a function $F(x)$ in the regions $x > 0$ and $x < 0$. In some sense this is contrary to the most familiar kind of polynomial approximation where, knowing how $F(x)$

behaves in the vicinity of a point $x = 0$ near which F is sufficiently smooth, we might attempt to model its variation in the form

$$F(x) \approx F(0) + xF'(x) + \tfrac{1}{2}x^2F''(0) + \dots , \tag{1.14}$$

with each higher order typically bringing a smaller correction to the model of $F(x)$. If instead $F(x)$ is discontinuous at a point $x = 0$, and an expansion like (1.14) is impossible, but we can instead base an approximation on an expansion of the general form (1.11).

The 'expansion variable' in (1.11) is a discontinuous quantity ν, which we call a *switching multiplier*. It has a dual identity, being a piecewise-constant 0 in $x < 0$ and 1 in $x > 0$, but taking values $\nu \in [0, 1]$ at the discontinuity. We capture this by writing

$$\nu = \text{step}(x) , \tag{1.15a}$$

where the right-hand side is essentially the Heaviside step function, but with a set of values at $x = 0$,

$$\text{step}(x) = \left\{ \begin{matrix} 1 \text{ if } x > 0 \\ 0 \text{ if } x < 0 \end{matrix} \right\} , \qquad \text{step}(0) \in [0, 1] . \tag{1.15b}$$

We will use the definition (1.15b) throughout this work. Despite writing (1.15a) we will not write ν explicitly as a function of x (i.e., we will not write it as $\nu(x)$), partly to avoid unnecessary clutter, but more importantly because this becomes inappropriate at the discontinuity, where x alone does not determine ν, and we will instead have to treat ν itself as a dynamic variable over the interval [0, 1].

The number of terms we retain in series approximations like (1.11) or (1.14) depend on how much information we possess to model F. In a series like (1.14), expanding to linear order in x is sufficient to model local behaviour around a zero $F(0) = 0$, while nonlinear terms are needed to model non-local behaviour or changes of stability. The series (1.11) is similar, in that expanding to linear order in ν is sufficient to model local zeros of F at the discontinuity, but nonlinear terms are needed to capture non-local phenomena and changes of stability. To understand why, we will first have to derive the induced dynamics that occurs as ν switches value across the discontinuity interval or 'layer' $0 \le \nu \le 1$. Even before we introduce this (in Chap. 5), we can show how nonlinear terms augment the range of behaviours possible in a model with a simple example.

Consider a particle that collides with a wall at some time $t = \tau$, switching its velocity from $u = 2$ to $u = -1$ in the process. This implies that the kinetic energy $K = \frac{1}{2}mu^2$ of the particle decreases from $K = 2m$ to $\frac{1}{2}m$ during the collision (the left graph in Fig. 1.1). Now let us assume that we can write $u = 2 - 3\nu$, where ν is a step function taking values $\nu = 0$ for $t < \tau$ and $\nu = 1$ for $t > \tau$, with $\nu \in [0, 1]$ at $t = \tau$. The quantity ν might be just a mathematical convenience, but it might also relate to some other physical quantity, such as an angular momentum imparted to the particle during collision, or a voltage induced by the collision if the wall is made of a dielectric elastomer. Substituting this expression for u into the kinetic energy now gives $K = \frac{1}{2}m(2-3v)^2$, but this function can explore a range of values $0 \le K \le 2m$ during the collision (see right graph in Fig. 1.1). The nonlinearity in v allows a larger range of energies to be modelled than the $\frac{1}{2}m \le K \le 2m$ suggested by the original model.

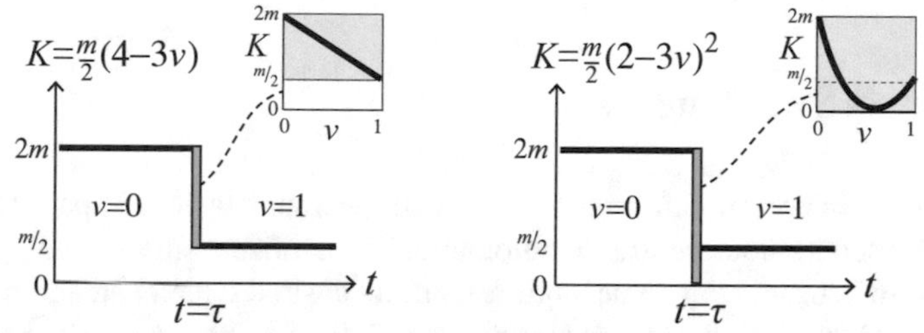

Fig. 1.1 Allowed values of kinetic energy through a discontinuity: linear (left) versus nonlinear (right) model. The upper right figures show the graph of K against v between 0 and 1, during collision

It is important to respect nonlinearity in discontinuous quantities, just as we respect nonlinearity in continuous variables. In the second model above the energy is given by $K = \frac{1}{2}m(2-3v)^2 = \frac{1}{2}m(4-3v+9v(v-1))$. Only the first part, $K = \frac{1}{2}m(4-3v)$, is visible outside the collision, while the nonlinear term $v(v-1)$ vanishes everywhere *except* at the collision. In other words the nonlinear term is *hidden* except where v transitions between 0 and 1.

The hidden term is able to sneak in to these equations because of the discontinuity at $t = \tau$. The loss of differentiability in Eqs. (1.1) invites similar hidden terms, see [61] for examples similar to (1a) and [81] for examples similar to (1b).

The issue of nonlinearity becomes more pressing in the context of multiple discontinuous quantities, say $v_1 = \text{step}(x_1)$ and $v_2 = \text{step}(x_2)$. The problem of handling multiple discontinuities has come to the fore particularly in models of genetic regulation, which we will discuss in Sect. 1.3, where discontinuous terms $v_i = \text{step}(x_i)$ can represent the action of numerous genes labelled $i = 1, \ldots, n$. Nonlinear phenomena then arise through multi-linear terms such as v_1v_2, for example, which appears indistinguishable from $\frac{1}{2}(v_1+v_2)$ in the regions where $v_1 = v_2$ with value 0 or 1, just as the term v_1^2 appears indistinguishable from v_1.

In the traditional literature on nonsmooth dynamics, the subject of nonlinear dependence on discontinuous quantities has received only brief attention, usually as counterexamples showing that nonlinear terms lie outside the theory that Filippov set out in [51]. In section 1.1. of [148], for example, Utkin shows that the system $\dot{x}_1 = \lambda x_1 + 0.3x_2$, $\dot{x}_2 = 4\lambda^3 x_1 - 0.7x_1$, with $\lambda = \text{sign}\,(x_1(x_1+x_2))$, displays different dynamics along $x_1 + x_2 = 0$ depending on whether one respects the λ^3 term as being distinct from λ. We explore a simplification of this example in Sect. 4.1.

We have argued here that if $\dot{x} = F$ and F is discontinuous, then (1.11) provides a series expansion in terms of a switching multiplier v. To assume that nonsmooth behaviour can be expressed in such closed form expressions as (1.11) is, however, still a huge idealization of the practical processes that typically accompany discontinuity, whether in electronics, mechanics, biological regulation, or numerous other disciplines where nonsmooth models are growing in prominence. We will therefore have three main themes in this article: to review the 'linear age' of nonsmooth dynamics in Chap. 2, to explore nonlinear or 'hidden' dynamics in Chap. 3

1.2 Control Switching

The principles of nonsmooth dynamics are well illustrated by a 1934 paper by Nikolsky, using discontinuous control to automatically stabilize a ship's heading [109].

If the ship is heading at some angle x from its desired course, an automatic control is applied that switches its rudder between left and right extremes, imparting a torque $\pm M$, illustrated in Fig. 1.2.

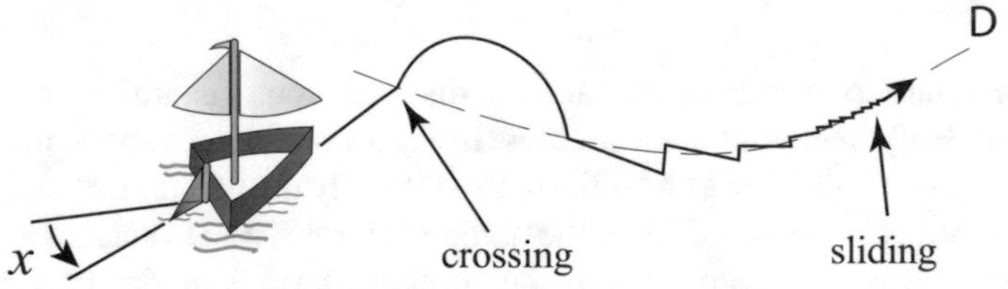

Fig. 1.2 Steering by discontinuous control, aiming along a heading $\mathcal{D}$

The equation of motion can be written as

$$I\ddot{x} = \nu M - G\dot{x} \, , \qquad \nu = \text{sign}(\sigma(x, \dot{x}, t)) \, , \tag{1.16}$$

where $I\ddot{x}$ is the angular acceleration and $G\dot{x}$ represents hydrodynamic damping. The control problem is essentially to design the function $\sigma(x, \dot{x}, t)$ that determines where switching occurs, with the aim of attaining the heading $x = 0$ as a stable dynamical behaviour.

In dynamical terms, the problem is simply to choose σ so that the system (1.16) has an attractor at $x = \dot{x} = 0$. Traditional dynamical systems theory, however, deals with differentiable systems, and can neither tell us how to solve (1.16) across $\sigma = 0$, nor how to understand its qualitative dynamics in terms of flows or attractors.

Yet what happens at $\sigma = 0$ is seemingly quite simple. From (1.16) we can see that with a large enough speed $|\dot{x}| \gg M/G$, the torque $I\ddot{x} = \pm M - G\dot{x}$ remains in the same direction as the $\pm$ sign switches. This describes what happens as we switch the rudder directly between its '$\pm$' settings. 'Solving' (1.16) is simply a matter of concatenating solutions of the two differentiable systems $I\ddot{x} = \pm M - G\dot{x}$ on the half spaces $\sigma > 0$ and $\sigma < 0$. We call this *crossing* of the discontinuity threshold $\sigma = 0$.

For small speeds $|\dot{x}| \ll M/G$, on the other hand, the torque $I\ddot{x} = \pm M - G\dot{x}$ changes sign as the ± sign switches. In this situation the rudder controller can begin chattering across $\sigma = 0$ between the '±' settings. If we can choose σ so that this chattering decays towards $\sigma = 0$, then the discontinuity set becomes attractive. If this leads to motion along a small neighbourhood of $\sigma = 0$ then it is known as *sliding* along the discontinuity; we will formalize the notion of sliding in Chap. 6.

Note that we have not yet even defined the switching function σ. An obvious choice is to take $\sigma = -x$, then the boat will always steer back towards $x = 0$ from either side. A more efficient solution is to let $\sigma = -x - \beta\dot{x}$ for $\beta > 0$, which anticipates that x will overshoot the desired heading based on the current rate of change $\dot{x}$, and so switches earlier.

Before we leave this example we can use it to illustrate some more novel features of nonsmooth dynamics. Simulating the flow in the $(x, \dot{x})$ phase plane using $\sigma = -x - \beta\dot{x}$, in Fig. 1.3(i) we see how the flow spirals in around a *fused focus*, until it hits the discontinuity threshold $\sigma = 0$, and then slides along the surface until reaching a fixed point at the origin (shown inset). This is what happens for $\beta > 0$. For $\beta = 0$ the size of the region of sliding shrinks to zero and solutions spiral in via infinitely many crossings. For $\beta < 0$ things are very different, as a *discontinuity-induced bifurcation* turns the fixed point unstable, and an attracting limit cycle grows out from the origin, reminiscent of an Andronov-Hopf bifurcation in differentiable systems. A central interest of nonsmooth dynamics over the last four decades has been the classification of fixed points and bifurcations like these, as can be found in [35, 51, 82, 93].

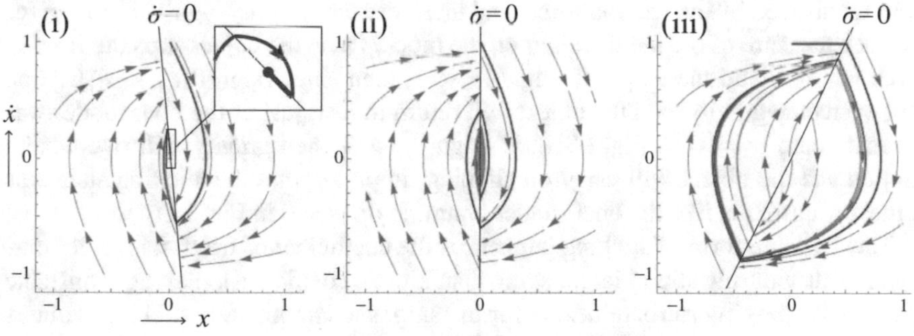

Fig. 1.3 Phase portrait of the boat steering control, taking the example $\sigma = -x - \beta\dot{x}$. (i) Shows a solution spiralling in and then sliding to fixed points, (ii) shows a solution spiralling in via infinitely many crossings, (ii) shows two solutions spiralling towards an attracting limit cycle. Simulated for $I = G = M = 1$, and β values (i) 0.1, (ii) 0, (iii) −0.5

We have skipped over how we handled the discontinuity to obtain the simulations in Fig. 1.3, and for good reason, as our aim here will be to bring to the fore various overlooked subtleties of such simulations. This problem brings us to consider the value of $\nu = \text{sign}(0)$ with care.

Letting $\sigma = -x - \beta\dot{x}$, there exists a special value, $\nu = \text{sign}(0) = (\beta G - I)\dot{x}/\beta M$, for which (1.16) re-arranges to $\beta\ddot{x} + \dot{x} = 0$, which implies $\dot{\sigma} = 0$. This means motion proceeds along $\sigma = 0$, so this is the value of $\nu = \text{sign}(0)$ that gives sliding dynamics. Note that this particular sliding solution cannot exist for $\beta = 0$ (because the proposed value of ν is singular), which is the case in Fig. 1.3(ii).

But is this the only possibility for the value of sign(0)? Let us say sliding does occur in Fig. 1.3(ii), i.e., that there exists motion along $\sigma = x = 0$ even if $\beta = 0$. Looking at Fig. 1.3(ii) we can immediately see that such motion would not be stable to any perturbation that kicked it away from $\sigma = 0$: the vector field would immediately carry it off around the fused focus. In physical terms, while on $\sigma = 0$ the rudder would seem to stick in its intermediate position, perhaps due to some mechanical jamming, despite the controller seemingly wanting to select one of its '±' positions (since the arrows in $\sigma \gtrless 0$ point unambiguously through $\sigma = 0$). If we tap the rudder out of its jammed sliding position, normal control (motion in $\sigma \gtrless 0$) will be resumed.

Anomalous behaviour like this brings us into the nonlinear age of nonsmooth systems theory because, as we shall see, nonlinear dependence on the discontinuous quantity ν makes such behaviour possible.

There is a simple table-top experiment you can do to find sign(0) in a similar problem involving friction. Place a small object on a surface that you can incline. If the object slips left or right along the surface it experiences a friction force $\text{sign}(\dot{x})F$, where $\dot{x}$ is its velocity relative to the surface and F is (approximately) constant. While the object sits at rest, stuck to the surface, the friction force is $\text{sign}(0)F$.

Now incline the surface as far as possible *just* before the object begins to slip, and tap the object gently down the slope. There is a downward gravitational force on the object, balanced by the friction force, and therefore the quantity $\text{sign}(0)F$ now gives us just a measure of the fixed weight on the block. Once the object slips the friction force will be F up the slope. If $|\,\text{sign}(0)| < 1$, then since $|\,\text{sign}(0)F| < F$ friction will overcome gravity, and the object will return to rest, just as the boat rudder was attracted onto $\sigma = 0$ in Fig. 1.3(i). If $|\,\text{sign}(0)| > 1$ then gravity will overwhelm friction and the object will carry on slipping, implying that the sticking state was somehow unstable, like the boat rudder jamming on $\sigma = 0$ in Fig. 1.3(ii).

This unstable form of sticking happens if the coefficient of friction is larger during stick than during slip (a larger static than kinetic coefficient), making it unstable to small 'tapping' perturbations. As for the ship, such anomalous sticking requires a nonlinear approach to modelling with discontinuities.

The concepts used at the end of the twentieth century to describe dynamics at a discontinuity bore remarkable similarity to those in Nikolsky's work and others of the time. The major change since then has been the growing introduction of ideas from singular perturbation theory, which we outline and extend in this article.

In this section we have seen the two key motivations for nonsmooth models over the past century: control switching and discontinuous contact forces. Much of the pre-1960s work on electromechanical control such as [7, 52, 92, 101, 109, 143] remained somewhat obscure until efforts to systematize the application to electronic control by Utkin in [146, 147, 148], and until Filippov's work to systematize the theory in [50, 51]. The kind of aeronautic control envisaged by Irmgard Flügge-Lotz in [52], for example, has found more practical aeronautical control applications such as [118, 127, 128, 153].

The application to modelling of contact forces in frictional stick-slip motion and impacts has both ensured the continued use of nonsmooth models, but perhaps also held back their advancement. Though the origins and properties of friction have been well understood through works like [19, 137], how to model dry-friction in dynamical models remains a subject of continuing study, see, for example, [13, 15, 26, 32, 72, 90, 115, 116, 121, 142, 151, 152] for a sample of just some of the various multiple scale and nonlinear factors involved. The kind of applications of interest are challenging in themselves, for example, in geological events such as earthquakes [113, 120], or ice flow [42, 126], or in automotive and industrial contexts of rotor vibration, gear rattle, brake wear, drill-string vibrations, see, e.g., [9, 10, 20, 30, 77, 85, 87, 105, 106, 119]. Particular challenges arise in systems with multiple points of contact, see, e.g., [9, 10, 22, 25, 27, 154]. Multiple points of contact create new issues because they create multiple sources of discontinuity, an issue we will visit several times from Chap. 4 onward.

1.3 Sticky Genes

Some of the most exciting applications opening up in nonsmooth dynamics involve living systems, from their social networks to their inner regulatory processes. An area particularly rich in models is that of gene regulatory networks, which are collections of molecular components, consisting of DNA, RNA, and proteins, that interact in an organic cell to direct growth and development of living organisms. Genes switch abruptly on or off, activating different processes via the expression of proteins.

A common model (see, e.g., [103, 117]) for the varying concentration x_i of a given protein labelled i is

$$\dot{x}_i = \mathcal{B}_i(\nu_1, \nu_2, \dots) - \gamma_i x_i \,, \qquad i = 1, 2, \dots \tag{1.17}$$

where γ_i is a decay constant, and $\mathcal{B}_i$ is a Boolean expression in terms of discontinuous ‘on/off’ quantities $\nu_i = 0$ or 1. Switching typically occurs as certain threshold values of protein concentrations, some $\sigma_i(x_1, x_2, \dots) = 0$, are reached.

To introduce some basic concepts, consider a trivially simple one-gene regulatory model with a protein concentration x varying as

$$\dot{x} = b\nu - x \,, \qquad \nu = \text{step}(x - \theta) \,. \tag{1.18}$$

The gene activation occurs at the threshold $x = \theta$. Let us assume $0 < b < \theta$. Then we have $\dot{x} < 0$ everywhere up to the discontinuity, suggesting that x tends to zero, its rate merely jumping from $\dot{x} = b - x$ to $\dot{x} = -x$ as x descends through $x = \theta$ (noting that $b - \theta$ is negative).

This behaviour is verified if we express ν as the limit of a smooth sigmoidal function, say

$$\nu = \lim_{p\to\infty} \frac{x^p}{\theta^p + x^p} . \tag{1.19}$$

This function is monotonic, ensuring at $x = \theta$ that $\dot{x} = b - \theta$ remains in the range $-\theta < b - \theta < 0$, and hence x is always decreasing.

We can easily obtain contradictory behaviour to this, however. Contrast this with a model that represents the discontinuity by a different limiting function, say $\dot{x} = b\hat{\nu} - x$, where

$$\hat{\nu} = \lim_{p\to\infty} \frac{x^p}{\theta^p + x^p} + \frac{c(\theta x)^p}{(\theta^p + x^p)^2} , \tag{1.20}$$

for some $k > g > 0$. This is purely for demonstration, but could, for example, model a burst of over-production of the protein at the moment switching takes place. Again assume $0 < b < \theta$. It is quite easy to show that $\dot{x}$ now becomes positive for x sufficiently close to θ, if $b/4\theta > c/(1+c)^2$. As x decreases through the threshold it then becomes stuck at $x = \theta r^{1/p}$, for some r that is a function of b, c, θ, satisfying $r^{1/p} \to 1$ as $p \to \infty$.

This is the same kind of anomalous sticking that we described in Sect. 1.2 due to a jammed rudder or static friction. Because we derived the discontinuous quantities from smooth functions in this case, we are able to see what causes the anomaly. Both models (1.19) and (1.20) are identical in the sense that $\nu = \text{step}(x - \theta)$ and $\hat{\nu} = \text{step}(x - \theta)$, for $x \neq \theta$ at least, but they differ in the value they assign to step(0) at the discontinuity threshold $x = \theta$. More precisely, if $0 \leq \nu \leq 1$ then at $x = \theta$ the switch in (1.19) permits only $-\theta \leq \dot{x} \leq b - \theta$, while (1.20) permits the wider range $-\theta \leq \dot{x} \leq \frac{b}{4c}(1+c)^2 - \theta$ if $b(1+c)^2 > 4c\theta$, similar to the collision models in Fig. 1.1 in Sect. 1.1.

Manipulating the expressions in (1.19) and (1.20) before taking the limit $p \to \infty$, we can show that the two switching rules are related by

$$\hat{\nu} = \nu + c\nu(1 - \nu) . \tag{1.21}$$

The last term vanishes when $\nu = 0$ or 1, i.e., it is a *hidden* term as introduced in Sect. 1.1, making the models indistinguishable for $x \neq \theta$. The behaviour it induces, such as the anomalous sticking above, is called *hidden dynamics*. Note that $\nu(1-\nu)$ is still of the Boolean form compatible with gene models, but describes a contribution turned 'off' in both states $x > 0$ and $x < 0$.

We see here that a nonlinear dependence on the discontinuous quantity ν produces non-trivial behaviour. Nonlinearity arises more naturally when we consider multi-gene systems where several Boolean multipliers $\nu_1, \nu_2, ...$ may multiply each other, each turning on/off the production of a protein $x_1, x_2, ...$ For instance take a growth rate $\dot{x}_1 = a + b\nu_1 + c\nu_2 + d\nu_1\nu_2$ dependent on discontinuous quantities $\nu_1 = \text{step}(x_1)$ and $\nu_2 = \text{step}(x_2)$. The rate coefficients b and c are activated across the sets $x_1 = 0$ and $x_2 = 0$. The coefficient d is only activated on the half-lines

$x_1 = 1 < x_2$ and $x_2 = 1 < x_1$, and its bilinear term $+dv_1v_2$ can result in non-trivial dynamics where the two thresholds intersect, at $x_1 = x_2 = 0$.

Two important elements to be explored in this article are suggested by the gene problem. One is how we model switches. In (1.19) and (1.20) they are the limit of smooth sigmoidal functions, but we must also consider switches that are the limit of other processes across the threshold $x = \theta$, involving factors such as small hysteresis or time delay or stochasticity. The robustness of nonsmooth models to perturbations of these kinds has been long discussed (see, e.g., [65, 123, 125]), but never clearly formulated, a gap we aim to fill here by formulating *implementations* of switching.

The second element is the nature of the dynamical object responsible for sliding/sticking on the discontinuity, namely whether it is a simple attractor, or a complex set exhibiting nonlinear behaviour. We will loosen the rigid standard notion of sliding/sticking considerably by looking afresh at some toy cases and applications.

As for friction, there is an accepted model for the physics behind gene activation, in this case the so-called Hill function, a sigmoidal curve that describes concentrations of substances during binding to receptors, originating from the study of oxygen molecules binding to haemoglobin [70]. Because activation typically occurs abruptly, the sigmoid Hill curve approaches a step function. There have been many interesting studies into the ramifications of nonsmooth behaviour in gene networks, particularly concerning sliding along, and attractors on, the activation thresholds that result from using either sigmoidal or step function models, for example, in [24, 33, 41, 88, 94, 103, 104, 117, 135], and many novel dynamical phenomena have been discovered. Nonsmooth models using step function are in many ways simpler, but have been hampered by a lack of theory for systems with multiple discontinuities, so in resolving issues of stability at the activation thresholds authors have tended to revert to sigmoidal functions. With a proper formulation we can observe all of the same behaviours whether we use discontinuous step functions or smooth sigmoids. Using discontinuous models simplifies the stability analysis, however, and helps in beginning to ask how robust these models are to the choice of activation function.

In living organisms the number of gene activation switches can number in the thousands, but is rarely known exactly, as discussed in [122]. Interest in large gene networks naturally tends to focus on statistical measures of dynamics across the network, and concern itself less with the novel dynamics that could be happening local to the various discontinuity sets.

1.4 The Plan

The rather basic notions explored in this section turn out to be important steps in probing the mathematical foundations of nonsmooth dynamics, leading us to ask how we describe the set $\sigma = 0$ as a mathematical object in order to model practicalities of switching in applications.

We will dig a little deeper into the history and key concepts of nonsmooth dynamics in Chap. 2. The difficulty of modelling practical nonsmooth behaviour, and its issues of uniqueness, are discussed informally in Chap. 3 before we set out in search of ways to characterize it mathematically.

In Chap. 4 we set out three paradigmatic systems which, though simple and ideal, exhibit complex dynamics revealing some novel problems of practical modelling of dynamics at discontinuities. We then turn to the different concepts of *sliding* required to understand these problems, setting out basic principles in Chap. 5 and extending some standard concepts in Chap. 6, in a way that makes sense of perturbations away from idealized discontinuity. In Chap. 7 these ideas are applied to resolve the ambiguities arising from Chap. 4.

The concepts from Chap. 6 open the door to countless novel phenomena that remain to be discovered, and we sketch out a few recent steps in this direction in Chap. 8. Some concluding remarks are made in Chap. 9.

Chapter 2
1930–2010: Nonsmooth Dynamics' Linear Age

By a nonsmooth system, we typically mean a system of ordinary differential equations in $\mathbf{x} \in \mathbb{R}^n$, dependent on some discontinuous parameter ν,

$$\dot{\mathbf{x}} = \mathbf{F}(\mathbf{x};\nu)\,, \qquad \nu = \mathrm{step}(\sigma(\mathbf{x}))\,, \tag{2.1}$$

or defined on disjoint regions,

$$\dot{\mathbf{x}} = \begin{cases} \mathbf{f}^1(\mathbf{x}) & \text{if } \sigma(\mathbf{x}) > 0\,, \\ \mathbf{f}^0(\mathbf{x}) & \text{if } \sigma(\mathbf{x}) < 0\,. \end{cases} \tag{2.2}$$

The first form is often associated with the application to electronic controllers by V. I. Utkin [146, 147], while the latter was tackled in a more general way by A. F. Filippov [50, 51] using differential inclusions. In fact Utkin and Filippov's works both mainly concern the same situation, as Utkin himself discussed recently in [149], namely when $\mathbf{F}$ is linear in ν such that

$$\dot{\mathbf{x}} = \mathbf{F}(\mathbf{x};\nu) = \nu\mathbf{f}^1(\mathbf{x}) + (1-\nu)\mathbf{f}^0(\mathbf{x})\,. \tag{2.3}$$

A discontinuity may also be introduced by applying some map

$$\mathbf{x} \mapsto \mathbf{R}(\mathbf{x}) \qquad \text{on } \sigma(\mathbf{x}) = 0\,, \tag{2.4}$$

at the discontinuity, for example, a restitution law $\dot{x} \mapsto -r\dot{x}$ during an impact (see, e.g., [35]), or a drop in mass $m \mapsto m/2$ during cellular mitosis (see, e.g., [43]).

With an explicit expression (2.3) in terms of a discontinuous term ν, Filippov showed that a very substantial theory of qualitative dynamics was possible. The partial solutions in $\sigma > 0$ or $\sigma < 0$ are described by standard dynamical systems theory, and the task of nonsmooth dynamics is to study the effect of concatenating those solutions at $\sigma = 0$, and to find any solutions that may evolve along $\sigma = 0$.

For solutions that traverse the discontinuity transversally there has been considerable progress in extending notions of stability from differentiable systems. To

M. R. Jeffrey, *Modeling with Nonsmooth Dynamics*,
Frontiers in Applied Dynamical Systems: Reviews and Tutorials 7,
https://doi.org/10.1007/978-3-030-35987-4_2

characterize the stability of orbits as they cross a discontinuity one may use the saltation matrix (see, e.g., [35, 97]), to describe bifurcations in an orbit's intersection with the discontinuity one has the discontinuity mappings (see [35, 110]), and to study the separation of orbits to establish connections between manifolds or existence of limit cycles one may extend the idea of Melnikov functions (see, e.g., [12, 62, 91], though at present there are large and increasing number of papers investigating this line of study). There is also work ongoing to extend other powerful tools, such as those of inverse integrating factors [23].

A particular area of interest has been the study of oscillations in the presence of impact or other discontinuous control actions. Two key problems concern the number of co-existing limit cycles that a given class of systems may contain, or studying how this number changes via global bifurcations.

The question of the number of limit cycles in a system follows in the spirit of Hilbert's 16th problem [69]. For nonsmooth systems this was dealt a decisive blow with the demonstration that even a piecewise-linear system can have arbitrarily many limit cycles, depending on the shape of the discontinuity threshold [99, 112]. Say $(\dot{x}, \dot{y}) = (-y, v)$, where $v = \text{sign}(\sigma(x, y))$. For $\sigma = x$ this system is a *fused centre*, similar to the phase portraits in Fig. 1.3 except every orbit is a closed cycle, so this has infinitely many periodic orbits, but they are not *limit* cycles. If we tilt this to $\sigma = x - y$, there are no closed cycles at all, similar to Fig. 1.3(i). But if we oscillate the discontinuity threshold by letting, say, $\sigma = x - \frac{1}{2}\sin(y)$, then every point $(x, y) = (0, n\pi)$ on $\sigma = 0$ for any integer n generates a limit cycle, in an alternating attracting (even n) repelling (odd n) sequence, and therefore infinitely many in number. A considerable literature exists restricting to simpler discontinuity thresholds, allowing multiple thresholds, or (to a lesser extent) allowing the vector fields to be nonlinear, with applications to electronic controllers and impact oscillators among the most studied, see, e.g., [29, 35, 150] for a more detailed review.

The bifurcation of limit cycles has a similarly rich literature. The most important realization has been that the 'hard' character of the discontinuity can actually be exploited to characterize *global* bifurcations by their *local* geometry near $\sigma = 0$. The study began in earnest with Feigin's study of C-bifurcations (the 'C' coming from the Russian word сшивать for solutions that 'sew' through a discontinuity threshold in [44, 45, 46, 47, 48]).

Figure 2.1(i) illustrates two distinct types of limit cycle, one which exists as a smooth cycle in the subsystem $\dot{\mathbf{x}} = \mathbf{f}^0(\mathbf{x})$, the other which passes transversally ('sews') through a discontinuity threshold $\sigma = 0$, with portions passing through both $\dot{\mathbf{x}} = \mathbf{f}^0(\mathbf{x})$ and $\dot{\mathbf{x}} = \mathbf{f}^1(\mathbf{x})$. The intermediary case is a smooth cycle in the subsystem $\dot{\mathbf{x}} = \mathbf{f}^0(\mathbf{x})$ that *grazes* the surface tangentially. Although the dynamics is global, the only substantive change occurs in the local geometry near grazing, Fig. 2.1(middle), which stretches trajectories apart according to a square root scaling. This creates a square root in the global return map to some section Π taken through the flow, Fig. 2.1(right). If an orbit gains or loses segments of sliding in such a bifurcation, then one obtains return maps that are locally piecewise linear or of power 3/2. These local forms are described by *discontinuity mappings*, and have been derived generally for the lowest codimension grazing bifurcations involving impact or sliding [21, 34, 35, 60, 110].

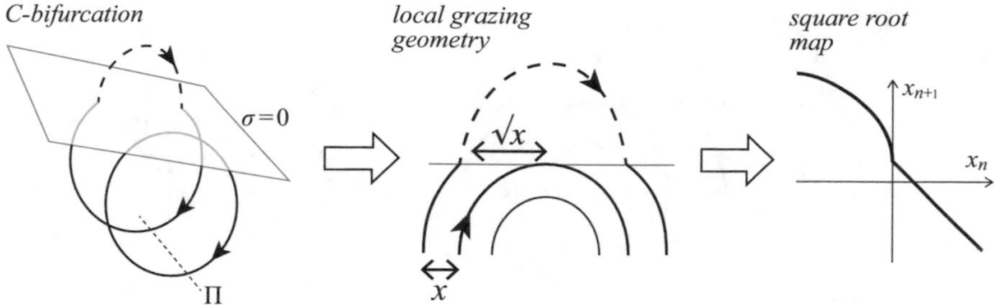

Fig. 2.1 A sewing or 'C' bifurcation, the local geometry near the discontinuity threshold, and the map induced on the return surface Π

The most well-studied nonsmooth maps are the square root map like that in Fig. 2.1(iii), piecewise-linear continuous maps $x \mapsto a + x(b + c\,\text{step}(x))$, and maps with a gap $x \mapsto a + bx + (1 + cx)\,\text{step}(x)$. The bifurcations undergone by their fixed points and periodic orbits exhibit considerable complexity, some key results of which can be traced through the papers [8, 11, 53, 56, 57, 73, 89, 100, 110, 114, 130, 133]. Most notable perhaps is the typical form taken by sequences of bifurcations of periodic orbits. For differentiable maps, an important result is the universality of the period doubling sequence by which a limit cycle bifurcates into eventual chaos, Fig. 2.2(i). By contrast, nonsmooth maps are almost unrestricted in the kinds of sequences they can exhibit, with the potential for periodic orbits to be appear of almost arbitrary period, or to jump suddenly to chaos. Figure 2.2(ii) shows a period incrementing sequence, interrupted by jumps to chaos, in a square root map.

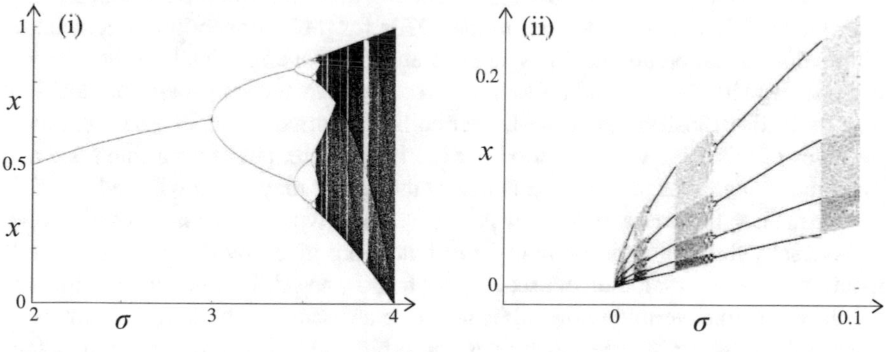

Fig. 2.2 (i) Period doubling cascade to chaos in a differentiable map, showing periodic attractors of the logistic map $x \mapsto \sigma x(1 - x)$. (ii) Period incrementing with windows of chaos in a nonsmooth map, showing periodic attractors of the square root map $x \mapsto \frac{3}{5}x + \sqrt{\sigma - x}\,\text{step}(\sigma - x)$ (from [35])

It remains an open problem to establish in general how nonsmooth flows and their global return maps are related, in particular how singularities and bifurcations of a grazing flow are related to the gradients, gaps, or power laws of return maps, especially in higher dimensions. As well as the power laws associated with grazing, it is known that sliding regions create horizontal or vertical branches in a map, as is illustrated in Fig. 2.3. If the discontinuity set is attracting, then many initial

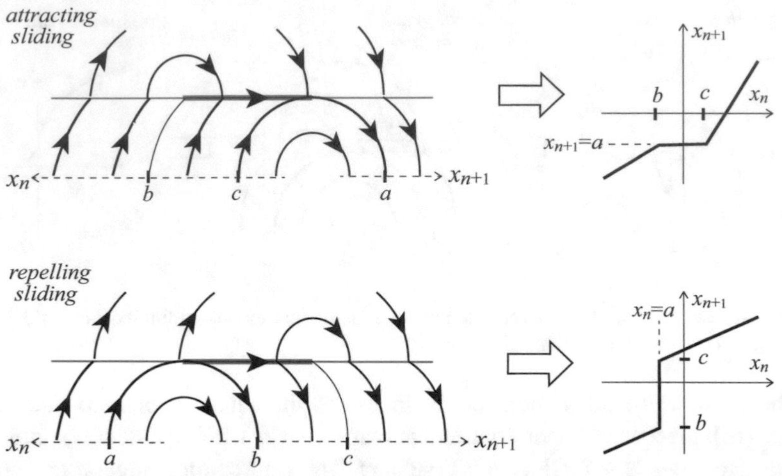

Fig. 2.3 Attracting or repelling sliding in a nonsmooth flow results in horizontal or vertical branches in return maps. Attracting: all points entering the depicted region with a coordinate $x_n \in [b, c]$ maps to an outgoing coordinate $x_{n+1} = a$. Repelling: a point entering the depicted region with a coordinate $x_n = a$ maps to a set of outgoing coordinates $x_{n+1} \in [b, c]$

conditions collapse onto the same sliding trajectory, resulting in a horizontal branch in a return map. Conversely if the discontinuity set is repelling, then any one initial condition in sliding explodes into a continuous family of trajectories outside sliding, resulting in a vertical branch in a return map.

The connection of nonsmooth maps to grazing flows has re-enlivened their study, after they received much attention in the 1970s for their connection to homoclinic bifurcations in differentiable flows, in generating robust chaos, and for their importance in ergodic theory, leading to such prototypes as the tent map, the doubling map or dyadic transformation, the border collision normal form, and the Lozi map. A survey of this topic would be too large to include here, but as a starting point and for some connections to more recent theory the reader may begin with [54, 59, 63].

The main tools necessary to study purely 'sewing' type behaviour—which evolves transversally through a discontinuity threshold—are given by the various methods mentioned above: a saltation matrix to describe how an orbit crosses a discontinuity, a Poincaré map describing how orbits return to a discontinuity, Melnikov methods to describe splitting between such returning orbits, and the discontinuity mappings associated with grazing. Henceforth we will be concerned mainly with far more troublesome issues related to the phenomenon of sliding, that is, evolution along discontinuities.

Filippov proposed in [51] that we study systems like (2.1) or (2.2) by forming a differential inclusion, making the system continuous but set-valued at $\sigma(\mathbf{x}) = 0$, by saying that $\dot{\mathbf{x}}$ lies in a connected set that contains the values of both $\mathbf{f}^0(\mathbf{x})$ and $\mathbf{f}^1(\mathbf{x})$ when $\sigma(\mathbf{x}) = 0$ (assuming the domains of $\mathbf{f}^0$ and $\mathbf{f}^1$ can be extended to such a point on the discontinuity threshold). By placing further assumptions of convexity and continuity on this set (section 4 of [51]), Filippov was able to prove that solutions existed, and even that they vary continuously with parameters and initial conditions.

Filippov's solutions, however, belonged to sets of *possible* solutions, with no criteria to select one solution over any other. He did offer one argument that provides unique solutions under certain conditions of stability, given by forming the differential inclusion from the smallest possible convex set that contains $\mathbf{f}^0(\mathbf{x})$ and $\mathbf{f}^1(\mathbf{x})$ (see section 4 part 1a of [51]). This is simply the family of vector fields generated by the right-hand side of (2.3) as ν varies over $\nu \in [0, 1]$, and constitutes the *linear* formulation of the nonsmooth problem.

Having a definite form (2.3) for the vector field allows us to solve for motion on $\sigma = 0$ as follows. At any $\mathbf{x}$ on $\sigma = 0$, if there exists some ν such that $\dot{\sigma} = 0$, then this gives *sliding* motion along $\sigma = 0$. Substituting that ν back into (2.3) gives the *sliding vector field*. If there exists no such ν, then crossing (or 'sewing') must occur. We already found such dynamics for the boat and genes of Sects. 1.2 and 1.3.

Being differentiable, the subsystems on $\sigma > 0$ and $\sigma < 0$ can exhibit equilibria within their respective domains. The sliding vector field on $\sigma = 0$ can also possess its own *sliding equilibria*. These have been given various names in different contexts, but I find them misleading in the wrong contexts. The term 'switched equilibria' is sometimes found, but we wish here to distinguish equilibria on the switching threshold that do or do not involve sliding. The term 'pseudo-equilibria' is very common, but 'pseudo' may suggest that they are somehow artificial, which they are not. Sliding equilibria are not, in particular, to be confused with 'virtual equilibria', which are points where $\mathbf{f}^1 = 0$ in $\sigma < 0$ or $\mathbf{f}^0 = 0$ in $\sigma > 0$, outside the domains of those vector fields as defined by (2.2)—these can influence dynamics via their drag on the nearby flow, but they do not exist as states of the system (2.2), and therefore are truly artificial. We shall use the term 'sliding equilibria' which more accurately identifies them with the sliding dynamics.

Bifurcations can occur in which equilibria become sliding equilibria or vice versa, or pairs of equilibria and sliding equilibria co-annihilate, known as boundary equilibrium bifurcations [35]. Partial classifications exist for planar systems [35, 51], but one must be careful, because in attempting to form such classifications it is easy to miss less intuitive cases like that in Fig. 2.4, see [74]. Boundary equilibrium bifurcations allow steady states of the system to be created on or off the discontinuity set, or to change stability, and in doing so create periodic orbits as we saw in Fig. 1.3.

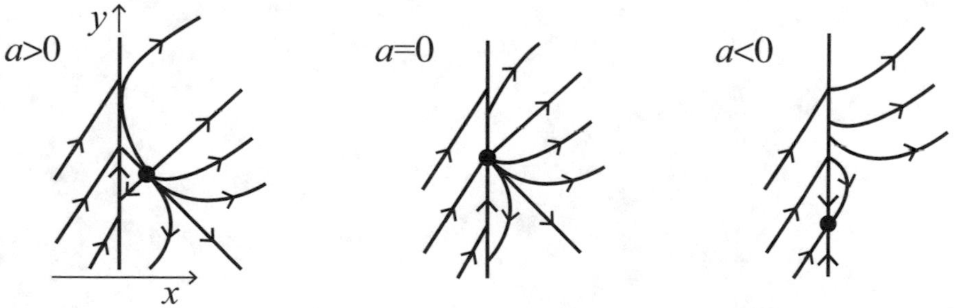

Fig. 2.4 A boundary equilibrium bifurcation in the system $\dot{x} = x+y-a-\nu$, $\dot{y} = y-2\nu$, $\nu = \text{step}(x)$. A repelling node in $x > 0$ becomes an attracting sliding node on $x = 0$ as a changes sign

There is much to be done in such systems classifying both local and global bifurcations, particularly in systems with multiple switches or more than two dimensions. For more detailed reviews of the state of the art in recent years the reader may also see, e.g., [29, 150]. But bifurcations, oscillations, and chaos are not the whole story. Like the dynamics of smooth systems around the middle of the twentieth century, nonsmooth dynamics is embarking upon its nonlinear age, with all the novel phenomena made possible by nonlinear dependence on discontinuous quantities like ν, and the difference between terms like $\nu, \nu^2, \nu^3, \ldots$ when $\nu = \text{step}(\sigma)$.

Chapter 3
Discontinuities to Model Missing Knowledge

If the theory of differentiable dynamical systems entered its nonlinear age with the discovery of bifurcations and chaos, then nonsmooth dynamics' nonlinear age will be characterized by tackling issues of determinacy. Loss of determinacy is an inescapable feature of nonsmooth systems, but it requires more explicit expression if it is to be turned to more useful modelling.

When integrating a function that involves something like a step function, the value step(0) at the discontinuity does not affect the value of the integral, provided the argument of the step function is monotonic over the integral path (see, e.g., chapter 1 of [51]). In a dynamic problem of the form $\frac{d}{dt}\mathbf{x} = \mathbf{F}(\mathbf{x}; \text{step}(t))$, for instance, solutions may be written as

$$\begin{aligned}\mathbf{x}(t) &= \mathbf{x}(\tau) + \int_\tau^t ds\, \mathbf{F}(\mathbf{x}(s); \text{step}(s)) \\ &= \mathbf{x}(\tau) + \int_\tau^0 ds\, \mathbf{F}(\mathbf{x}(s); 0) + \int_0^t ds\, \mathbf{F}(\mathbf{x}(s); 1)\,,\end{aligned}$$

assuming $\tau < 0 < t$, whose existence is provided by Carathéodory's theorem (extending Peano's existence theorem to non-differentiable systems of ordinary differential equations, see, e.g., [28, 51, 67]). The step function simply divides the integral in two, and the value of step(0) does not affect the right-hand side. In a dynamic problem of the form $\dot{\mathbf{x}} = \mathbf{F}(\mathbf{x}; \text{step}(\sigma(\mathbf{x})))$ for some function σ, however, where the discontinuity threshold is the set $\mathcal{D} = \{\mathbf{x} : \sigma(\mathbf{x}) = 0\}$, when seeking solutions of the Carathéodory form,

$$\mathbf{x}(t) = \mathbf{x}(a) + \int_a^t ds\, \mathbf{F}\Big(\mathbf{x}(s); \text{step}\,(\sigma(\mathbf{x}(s)))\Big)\,,$$

the value of the step function becomes essential, because the argument $\sigma(\mathbf{x}(s))$ could remain at zero for a non-vanishing interval of s values. The *existence* of sets of Carathéodory type solutions to problems of the form (2.2) was covered rigorously

M. R. Jeffrey, *Modeling with Nonsmooth Dynamics*,
Frontiers in Applied Dynamical Systems: Reviews and Tutorials 7,
https://doi.org/10.1007/978-3-030-35987-4_3

in [51], but in order to use those solutions for the purposes of modelling, we require a more explicit way to characterize them, to distinguish the different dynamics they make possible.

Mixed up in this problem is the understanding of what 'nonsmoothness' represents as an approximation. Nonsmooth systems are idealized in the sense that they take abrupt transitions, and represent them as discontinuities occurring at definite hypersurfaces in space. In what applied contexts is this a suitable model of abrupt change, and what physical or living processes can it faithfully represent? What are the implications of regularizing a discontinuity to obtain a well-defined system, and does it matter *how* we regularize? Should we obtain similar behaviour whether we smooth the discontinuity out, or numerically calculate the state $x(t)$ by interpolating between time intervals $t, t+\delta_1, t+\delta_2, \ldots$? What happens if the true system actually lies a short distance δx away, or suffers a time lag δt, shifting $x(t)$ to $x(t-\delta t)+\delta x(t-\delta t)$, and does it matter whether these perturbations are simple functions or stochastic processes? Should all such implementations of switching give similar *perturbations* of the nonsmooth model, and if not, how do we make sense of them as approximations of some underlying physics? Various such problems have been discussed in the literature in a somewhat fragmented manner. Filippov's formulation in [51] used differential inclusions, while a recent trend to smooth the discontinuity has followed the approach introduced in [136] (and extended to include hidden dynamics in [31, 111]). The discontinuity's implementation is considered with spatio-temporal delays or 'chatter' in [2, 6, 129, 148], or with stochasticity in [132].

Part of the difficulty of developing a theory of how discontinuities affect dynamics may be in our lack of understanding of what causes them. They tend to be used when we have incomplete knowledge of the underlying processes when some abrupt transition occurs, so we apply different set of equations under different conditions, in different regions of state or parameter space. A way to clarify what a discontinuity represents in a model might be to separate out those that are *passive*, *active*, or *dynamic* in origin.

A *passive* discontinuity is simply a change in the physical parameters defining a system, such as density, conductivity, or reflectivity, which jump across the interface between different materials. A discontinuity in refractive index, for example, is responsible for light bending as it passes between air and water. The jump in density between air and metal allows a hammer to swing freely through the air, and yet impart a force when it contacts with a nail. To precisely understand the frictional and impact contact forces between such media requires a microscale understanding of their interfaces, but such a level of detail can hardly be useful in a large scale model of the dynamics of the bodies themselves, so we approximate using, among other simplifications, discontinuities.

An *active* discontinuity is imposed upon a system as a means of control. Examples might be switches or valves made of mechanical, electronic, biological, or chemical parts, opening and closing different channels that drive or starve different parts of a system. Or they may be decisions made by individuals or groups about how to govern institutions, how to invest, or what causes to support. How we model such discontinuities depends on whether they are imposed instantaneously, gradually, or via a series of sub-processes. With potentially so many complex factors involved, the more detailed the model the less general would be any results obtained from it.

A *dynamic* discontinuity is induced by some small and/or fast scale change in stability, a jump from one stable attractor or pattern to another. The transition may be too abrupt to be easily observed itself, but manifests as a large scale change in behaviour, such as a financial crash, a change in heart rhythm, the collapse of a structure, or onset of turbulence. Of our three categories, these have the most comprehensible origins in the form of bifurcations or phase transitions, for which we have well developed mathematical theory, such as the asymptotic theory discussed in Sect. 1.1, particularly as concerns Stokes discontinuities and shocks [14, 16, 39, 68, 71].

These distinctions are not definitive, nor are they unique. For example, if a person makes a decision *actively*, then this may actually be the result of some *dynamic* phase transition across networks of neurons in the brain. But they illustrate the different features that discontinuities are called upon to approximate, the huge complexity that is disguised by writing $|x|$, step(x), or other conditional statements, in a systems of differential equations. We must formulate nonsmooth dynamics in a way that is explicit enough to explore the assumptions behind using such terms, and robust enough to start relaxing those assumptions.

Such issues were already in the minds of the pioneers of nonsmooth theory. It is worth recounting Filippov's own thoughts from section 8 of [51] on what nonsmooth models are and what they aspire to:

1. Differential equations with discontinuous right-hand sides are often used as a simplified mathematical description of some physical systems. The choice of one or another way of definition of the right-hand side of the equation on a surface of discontinuity ... depends on the character of the motion of the physical system near this surface.
2. Suppose that outside a certain neighbourhood of a surface of discontinuity of the function $F(t, x)$ the motion obeys the equation $\dot{x} = F(t, x)$. In this neighbourhood the law of motion may not be completely known. Suppose the motion in this neighbourhood may proceed only in two regimes, and switching over from one regime to the other has a retardation, the value of which is known only to be small.
3. Using these incomplete data, we should choose the way of defining the right-hand side of the equation of the surface of discontinuity, so that a sufficiently small width of the neighbourhood the motions of the physical system differ arbitrarily little from the solutions of the equation $\dot{x} = F(t, x)$ defined in the way we have chosen[1].

The subtle problem Filippov himself poses here has been largely overlooked in the advancing theory of nonsmooth dynamics. Here we will loosen the main concepts of nonsmooth dynamics in a way that allows us to probe these issues more deeply.

[1]In the original text this appears as one continuous passage, but for emphasis we have broken it into three items.

If discontinuities were all known to be of the dynamic kind, then nonsmooth dynamics would be just a direct extension of nonlinear dynamics and singular perturbations. Instead, the dominant influence in the development of nonsmooth dynamics has been of the *active* kind, from electronic and mechanical control, through to biological regulation, where we often have incomplete knowledge of the processes behind the discontinuity. In the next two sections we use these applications to introduce some basic concepts.

A lack of a way to quantify and ultimately reconcile such different approaches to handling discontinuity remains the 'elephant in the room' for nonsmooth dynamicists. Nevertheless, given the successes of these different lines of study we should now be able to form a more general picture. To do so we shall have to slightly loosen some of the standard concepts of nonsmooth dynamics. What we will show here is that the different behaviours that are possible at a discontinuity can be distinguished explicitly. We will also show that in regularizing the discontinuity one may unwittingly single out only one of the many possible behaviours, such that different treatments of the discontinuity giving contradictory dynamics.

The path to a general theory is laid by forming a common framework to describe how solutions 'handle' or *implement* a discontinuity (less restrictive than attempting to 'regularize' the discontinuity), along with recognizing a common set of behaviours that such implementations give rise to.

Our first steps towards formalizing such a framework here culminate in a conjecture that can be paraphrased as:

> solutions of a system of ordinary differential equations with a discontinuity along a threshold $\mathcal{D}$, lie ε-close to solutions of a similar system in which discontinuity occurs in some region $\mathcal{D}^{\varepsilon}$, such that $\mathcal{D}^{\varepsilon} \to \mathcal{D}$ as $\varepsilon \to 0$.

Such a conjecture cannot be expected to hold without adequately specifying how switching takes place across $\mathcal{D}^{\varepsilon}$ between differentiable regimes outside $\mathcal{D}^{\varepsilon}$. Filippov's own work (see, e.g., sections 8–12 of [51]) considered ε-perturbations away from the ideal discontinuous equations to prove the existence of solutions, and that they travel along the convex hull of the nearby vector fields (we will look at this hull in Sect. 5.1), leading to sets or 'funnels' of possible solutions. Here we will look at how to model specific vector fields and solutions among those possibilities, to find to what extent we can obtain a well-defined and robust model. We shall show that non-idealities, no matter how small, play a crucial and fascinating role.

Chapter 4
Three Experiments

To set up the remainder of this article let us pose three problems. They incorporate various issues arising from basic analysis to simulation to applied modelling. Their seemingly ambiguous dynamics expose remarkably well our current state of knowledge concerning the robustness of nonsmooth models. The problems are: a classic example of ambiguity from seminal texts; a two-gene regulatory system with seemingly ambiguous activation of genes; and an investment game where players' seemingly steady behaviour destabilizes a company's trading.

To solve these systems we must decide how to handle a discontinuous term like step(σ), in particular, to decide what value to assign to step(0). We can do this in a number of different ways, which we call *implementations*. We will make a preliminary definition of these here, to be refined in Chap. 5, along with six *test implementations* that we use for simulations in this article.

Definition 4.1. Given a system in which a parameter switches according to an ideal rule ν = step(σ), we define an **implementation** of the switch as a rule that forms a transition layer $|\sigma| \leq \varepsilon$ and determines the selection of modes $\nu = 0$ and $\nu = 1$ within it. In $|\sigma| > \varepsilon$ the modes are assigned uniquely according to ν = step(σ). For our purposes here this can involve one or more of the following six *test implementations*:

- *smoothing*—the switch is enacted by replacing ν with a smooth monotonic function $\phi^\varepsilon(\sigma)$ satisfying $\phi^\varepsilon(\sigma) \to$ step(σ) as $\varepsilon \to 0$;
- *hysteresis*—the switch is enacted by mapping $\nu : \{0\ or\ 1\} \mapsto 1$ at $\sigma = +\varepsilon$, and $\nu : \{0\ or\ 1\} \mapsto 0$ at $\sigma = -\varepsilon$;
- *time delay*—the switch is enacted by mapping $\nu : 0 \leftrightarrow 1$ a time $\delta t = \varepsilon$ after crossing $\sigma = 0$, that is, ν = step($\sigma(t - \varepsilon)$);
- *time stepping*—the rule ν = step(σ) is evaluated only at discrete time steps $\delta t = \varepsilon$;
- *noisy time delay*—as in time delay, but with the delay chosen randomly between $0 < \delta t < \varepsilon$ each time a switch occurs;
- *spatial noise*—as in smoothing, but the state is also perturbed by adding random translations of a size smaller than some $\kappa \ll 1$ at time steps $\delta t = \alpha \ll 1$.

M. R. Jeffrey, *Modeling with Nonsmooth Dynamics*,
Frontiers in Applied Dynamical Systems: Reviews and Tutorials 7,
https://doi.org/10.1007/978-3-030-35987-4_4

In each case ε is a small positive constant. To consider a system with multiple switches we introduce parameters $\nu_i = \text{step}(\sigma_i)$ for $i = 1, 2, ...,$ which switch at thresholds $\sigma_i = 0$, and in the case of smoothing, hysteresis, or time delay, involve different small quantities ε_i, $i = 1, 2, \ldots$ associated with each threshold.

We have expressed the switch in terms of $\nu = \text{step}(\sigma)$ without loss of generality, and could instead have used any other discontinuous function, such as $\lambda = \text{sign}(\sigma)$ for which the implementations follow directly via the relation $\lambda = 2\nu - 1$.

Of course one may consider other implementations besides those above, for instance perturbations that increase the dimension of the system. The list above will be sufficient for this exploratory study, but there are certainly extensions to be made, and perhaps a more general but less explicit formulation will be found. We will not consider the implementation that is provided by using event detection in computational software, because this requires an additional decision to be made of what dynamics to impose on $\sigma = 0$, a decision we are not yet ready to make.

4.1 Filippov's Convexity Paradox

The following problem is adapted from one posed by Filippov in section 7 example 134 of [51]. Consider a planar system

$$\begin{pmatrix} \dot{x} \\ \dot{y} \end{pmatrix} = \begin{pmatrix} -\frac{1}{2} - \lambda \\ \frac{1}{3} + \lambda^3 \end{pmatrix}, \tag{4.1}$$

where $\lambda = \text{sign}(x)$, and assuming $\lambda \in [-1, +1]$ when $x = 0$. Whatever the flow in this system, the vector field for $x \neq 0$ simply points towards $x = 0$, so the flow is attracted to the discontinuity at $x = 0$ and must remain there. All that remains is to find its speed of motion along $x = 0$.

In Fig. 4.1(i–vi) we simulate (4.1) using the test implementations of the switch set out in Definition 4.1. Smoothing the discontinuity (and then simulating using some standard numerical integration package), we obtain the trajectory in portrait (i) of Fig. 4.1. Simply discretizing Eqs. (4.1) instead gives the portrait (ii). Introducing hysteresis or delay in the switch yields similarly (iii)–(iv), while a noisy delay yields (v). In (vi) we see two examples of what happens with a combination of such effects, as we smooth Eqs. eq. (4.1), discretize them, and then introduce spatial noise: for small enough noise the simulation agrees with (ii–v), for large noise it agrees with (i), up to random variations. The sizes of these perturbations are described in the figure caption.

All show motion along the discontinuity threshold, but some result in motion upward at a speed $\dot{y} \approx 5/24$, while others evolve downward at a speed $\dot{y} \approx -1/6$ (we will see why these specific values of the speed arise). Implemented with noise, either type of solution can be obtained for sufficiently large or small amplitude of random perturbations.

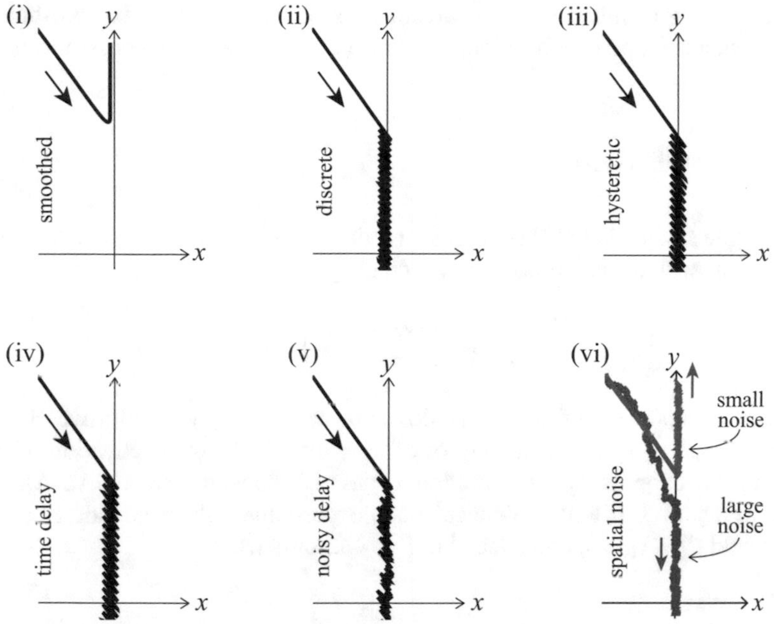

Fig. 4.1 The system from Fig. 7.1 simulated from a point in $x < 0$, where the switch implementation is: (i) smooth, (ii) time stepping, (iii) hysteretic, (iv) time-delayed, (v) with a noisy time delay, or (vi) with spatial noise (large or small noise, slightly displaced for clarity). Simulated using (4.1) and Definition 4.1 with $\varepsilon = 0.1$. In (v) we use $\kappa = \alpha = 0.1$ for small noise and $\kappa = 0.2$, $\alpha = 0.01$, for large noise. Similar results are obtained for any initial conditions and any choice of smoothing function

This was essentially this problem posed by Filippov, but considering only an idealized system (i.e., not considering such implementations), to highlight issues of uniqueness of solutions to (4.1). Filippov's theory, taken as a whole, allows for all of the different outcomes in Fig. 4.1, but his most commonly accepted concept of 'sliding' along the discontinuity threshold favours the outcomes in (ii)–(v), in contradiction to the outcome of smoothing in (i).

Given such a simple system as (4.1), the contrasting results in Fig. 4.1 reveal somewhat of a conceptual pickle in the state of understanding of nonsmooth dynamical theory, one that will become more pronounced over the next couple of examples. Distinguishing between the different possibilities and when to expect them is vital for applications, and moreover is achievable given recent advances, such as [18, 79, 83, 129, 131, 132]. These have provided hints of conditions to distinguish between alternative behaviours, but rigorous results are piecemeal, lengthy, and technical, even in the simplest cases when one smooths a discontinuity [111] or adds hysteresis [18].

One might be concerned that we can hope at all to develop a serious theory of nonsmooth models in the light of such contrary results. We will show that these are each correct and reconcilable within definite perturbative contexts. We will also show that these pale against other indeterminacies that have come to light in recent years, not limited to speed or direction of motion along the threshold, but also whether solutions evolve along the threshold at all or simply cross through it.

4.2 On or Off Genes

An example studied in [117] concerns a model of two genes responsible for producing proteins with concentrations x_i, which obey

$$\begin{pmatrix} \dot{x}_1 \\ \dot{x}_2 \end{pmatrix} = \begin{pmatrix} \nu_1 + \nu_2 - 2\nu_1\nu_2 - \gamma_1 x_1 \\ 1 - \nu_1\nu_2 - \gamma_2 x_2 \end{pmatrix} := \mathbf{f}^{\nu_1\nu_2}(x_1, x_2), \tag{4.2}$$

where $\nu_i = \text{step}(x_i - \theta_i)$ for some positive constants $\theta_1, \theta_2, \gamma_1, \gamma_2$, for which we assume $0 < \gamma_i\theta_i < 1$. The switching of each ν_i models the (de-)activation of the ith protein production as its concentration x_i passes a threshold value θ_i, as we introduced in Sect. 1.3. It will be convenient to express the right-hand side as a planar vector field $\mathbf{f}^{\nu_1\nu_2}(x_1, x_2)$, as sketched in Fig. 4.2(top-left).

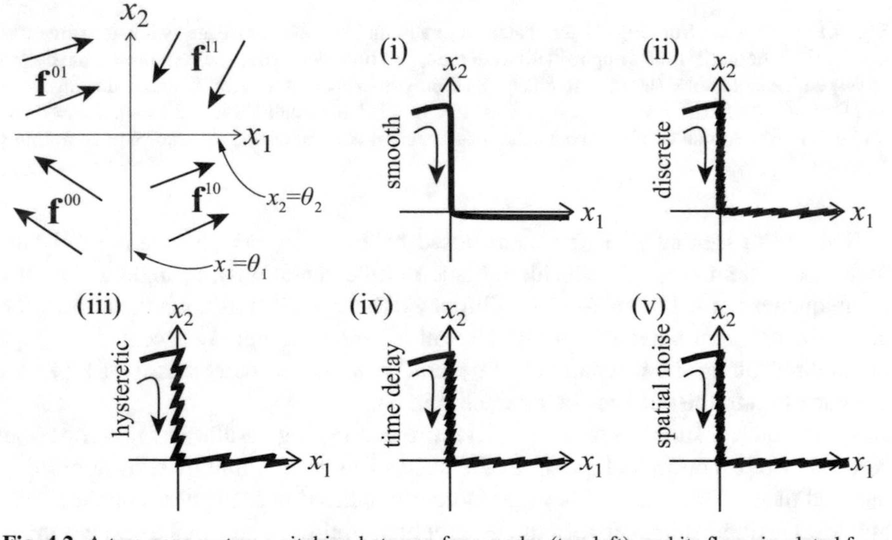

Fig. 4.2 A two-gene system switching between four modes (top-left), and its flow simulated from a point in $x_1 < \theta_1$, $x_2 < \theta_2$, in (i–v). The solution evolves onto the discontinuity threshold $x_1 = \theta_1$, and proceeds onto the threshold $x_2 = \theta_2$, where it then remains approximately. The switch is modelled as: (i) smooth (substituting ν_i with $\nu_i = x_i^{1/\varepsilon_i}/(\theta_i^{1/\varepsilon_i} + x_i^{1/\varepsilon_i})$ for small constants ε_i), (ii) time stepping, (iii) hysteretic, (iv) time-delayed, or (v) with a noisy time delay. Simulated from (4.2) for parameters $\theta_1 = \theta_2 = 1$, $\gamma_2 = 0.9$, $\gamma_1 = 0.4$, and implemented with $\varepsilon_1 = \varepsilon_2 = \alpha = \kappa = 0.2$ (as described in Definition 4.1)

The range of behaviours that are possible at the discontinuity thresholds is less obvious than in Sect. 4.1. Clearly the half-lines $x_1 = \theta_1$, $x_2 > \theta_2$, and $x_2 = \theta_2$, $x_1 > \theta_1$ are attracting, so motion can only proceed along the thresholds there, and solutions can only escape those portions at the point $x_1 - \theta_1 = x_2 - \theta_2 = 0$ where they intersect. Elsewhere, on $x_1 = \theta_1$, $x_2 > \theta_2$, and $x_2 = \theta_2$, $x_1 > \theta_1$, it can be shown that solutions cross these thresholds transversally (see [82, 117]). Let us focus solely on what happens at $x_1 - \theta_1 = x_2 - \theta_2 = 0$ after motion along $x_1 - \theta_1 = 0 < x_2 - \theta_2$.

Similarly to Sect. 4.1 we simulate the system by implementing the two discontinuities in the manner of Definition 4.1. The results are shown in Fig. 4.2 for parameter values given in the caption. In all cases the solution evolves onto the threshold $x_1 = \theta_1$, $x_2 > \theta_2$, travels downward until reaching the intersection $x_1 - \theta_1 = x_2 - \theta_2 = 0$, and 'rounds the corner' onto the threshold $x_2 = \theta_2$, $x_1 > \theta_1$. (In fact the solution continues towards an equilibrium state at $x_1 \approx \theta_2\gamma_2/\gamma_1$, $x_2 \approx \theta_2$, not shown). The outcome appears to be robust to the method of implementation.

For a different set of parameters, however, we obtain the contradictory behaviours shown in Fig. 4.3. This is despite the vector fields outside the discontinuity thresholds $x_i = \theta_i$ being qualitatively unchanged. Implemented by smoothing, time stepping, or time delay, the solution now sticks upon reaching $x_1 - \theta_1 = x_2 - \theta_2 = 0$. With hysteresis the solution rounds the corner as before; however, isolated param-

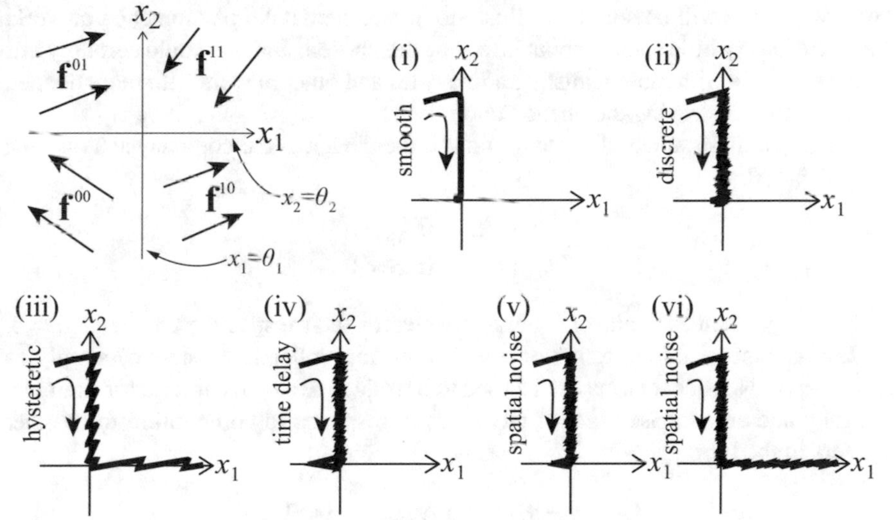

Fig. 4.3 As in Fig. 4.2 except $\gamma_1 = 0.6$. The solution evolves onto the discontinuity threshold $x_1 = \theta_1$, along which it evolves onto $x_1 = \theta_1$, $x_2 = \theta_2$, where it then remains (approximately), for all implementations except the last—noisy solutions can either stick (v) for $\kappa \gg 0.02$, or round the corner (vi) for $\kappa \ll 0.02$, while near $\kappa \approx 0.02$ intermittency is seen, with repeated simulations showing either behaviour

eter values $\varepsilon_1 \neq \varepsilon_2$ can be found for which the hysteretic solution does stick to $x_1 - \theta_1 \approx x_2 - \theta_2 \approx 0$, for example, $\varepsilon_1 = 0.01$, $\varepsilon_2 = 0.02$ (not shown). Implementing with noise we are able to see both behaviours: sticking at the origin for small enough noise, rounding the corner for larger noise, and in between, an intermittency whereby repeated simulations may show either behaviour due to the randomness of perturbations.

These contrary results can be explained, and a context to do so is provided in the following sections. In short they hinge on the existence or not of an attractor at $x_1-\theta_1 = x_2-\theta_2 = 0$ and its stability under perturbation. Re-considering the literature in the light of the results above, one recognizes hints towards them in studies such as [117] and [2]. In [117] the dynamics at the threshold is studied assuming the quantities v_i are actually smooth 'Hill' functions, $\phi_i^{\varepsilon_i}(x_i) = x_i^{1/\varepsilon_i}/(\theta_i^{1/\varepsilon_i}+x_i^{1/\varepsilon_i})$, which tend to step$(x_i - \theta_i)$ as $\varepsilon_i \to 0$ [70].

Even if solutions stick robustly to the intersection $x_1 - \theta_1 = x_2 - \theta_2 = 0$, their fate might still not be obvious, as the following example shows.

4.3 Jittery Investments

Consider a game in which two players each buy and sell stocks in a company, denoting the amounts they own as x_1 and x_2, measured relative to some desirable asset level $x_i = 0$, and let the stocks owned by the company itself be z. The system and phenomenon we describe below extend readily to more players and companies. The rules we assign will be simple to illustrate the mathematical phenomenon at work, but there is very little special about how they are chosen, and one could certainly improve the game with more realistic trading rules and other players with more defined roles without destroying the phenomenon.

Let x_i purchase stocks at a rate c, and sell them back to the company at a rate $-\gamma_i$ only if $x_i > 0$, so

$$\dot{x}_i = \begin{cases} c - \gamma_i & \text{if } x_i > 0 \, , \\ c & \text{if } x_i < 0 \, . \end{cases}$$

If $0 < c < \gamma_i$, then each player is attracted towards their respective threshold $x_i = 0$.

Let us express this in terms of a switching multiplier $v_i = \text{step}(x_i)$ simply as $\dot{x}_i = c - \gamma_i v_i$. Nonlinear terms can be used to introduce strategic choices, for example, we may add an increased rate of buying ρ_i in response to competition from other players in the form

$$\dot{x}_i = c - \gamma_i v_i - \rho_i v_1 v_2 \, , \qquad i = 1, 2. \tag{4.3}$$

The company's level of self-owned stocks is then fed and depleted by the two players, giving

$$\dot{z} = -2c + \gamma_1 v_1 + \gamma_2 v_2 - \rho_{11} v_1 v_2 + \rho_{00}(1 - v_1)(1 - v_2) \, , \tag{4.4}$$

where we also give the company its own strategic terms, namely an increase in selling off to other parties at rate ρ_{11} if both players are selling at the same time, or an increase in buying back shares from other parties at rate ρ_{00} if both players are buying at the same time.

We will denote the right-hand side of the three-dimensional vector field defined by (4.3)–(4.4) as $\mathbf{f}^{\nu_1\nu_2}$, so

$$(\dot{x}_1, \dot{x}_2, \dot{z}) = \mathbf{f}^{\nu_1\nu_2} \tag{4.5}$$
$$:= \Big(c - \gamma_1\nu_1 - \rho_1\nu_1\nu_2\,,\ \ c - \gamma_2\nu_2 - \rho_2\nu_1\nu_2\,,$$
$$-2c + \gamma_1\nu_1 + \gamma_2\nu_2 - \ \rho_{11}\nu_1\nu_2 + \rho_{00}(1-\nu_1)(1-\nu_2) \Big)\,,$$

and we assume $\gamma_i + \rho_i > c > 0$ for $i = 1, 2$.

The piecewise-constant vector field is sketched in the (x_1, x_2) plane in Fig. 4.4, and again appears extremely simple: the flow evolves onto the thresholds $x_1 = 0$ and $x_2 = 0$, where it must remain. The simulations in Fig. 4.4 reveal that for any implementation (i)–(v), solutions evolve towards the intersection $x_1 = x_2 = 0$ and remain there for all later times.

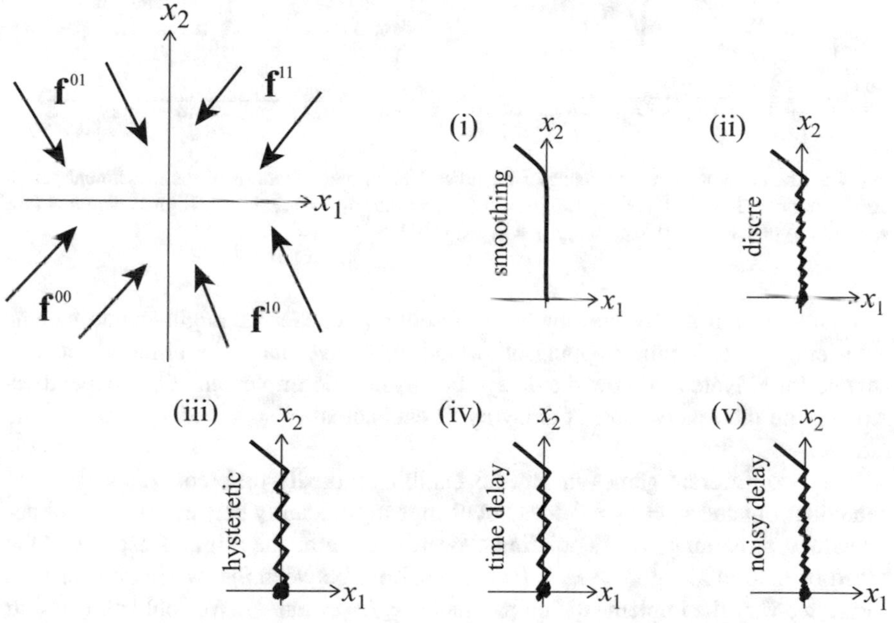

Fig. 4.4 An investment game with four modes. The vector field is sketched (top-left), and its flow is simulated in (i–v) from a point in $x_1 < 0 < x_2$ by the different implementations from Definition 4.1. The solution evolves onto the discontinuity threshold $x_1 = 0$, then towards the intersection with the second discontinuity threshold $x_2 = 0$, where it remains (approximately), regardless of implementation, and for any parameter values (given $\gamma_i + \rho_i > c > 0$ for $i = 1, 2$)

Thereafter only the company's holdings, z, may fluctuate, so we must ask how this is affected by the investment dynamics of the two players. As in Sects. 4.1 and

4.2, our task reduces to determining the speed of motion along the discontinuity threshold, in this case the rate of change of z on $x_1 = x_2 = 0$.

We find a rather different phenomenon, with more erratic variability, than in Sects. 4.1 and 4.2.

The result of simulations for the implementations using smoothing, hysteresis, time delay, or time stepping is shown in Fig. 4.5, for two different games (i.e., two sets of parameters). The speed $\dot{z}$ is plotted for different ratios of the implementation parameters ε_i, which are defined as $(\varepsilon_1, \varepsilon_2) = \varepsilon(\cos\frac{\omega\pi}{2}, \sin\frac{\omega\pi}{2})$. Note for time stepping there is only one ε parameter (the time step) so the graph is a horizontal line.

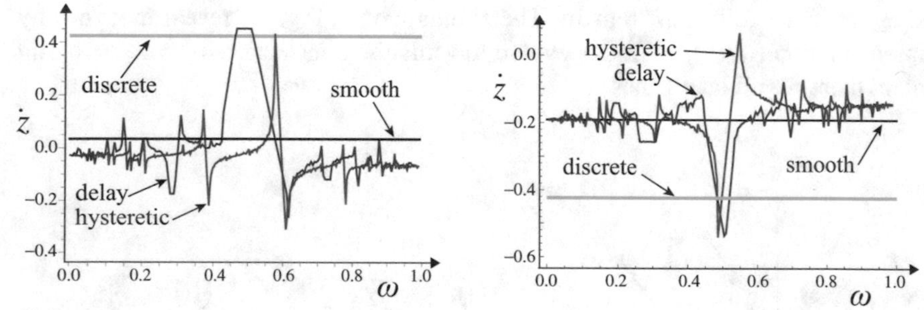

Fig. 4.5 The company's rate of change $\dot{z}$ for noise-free implementations of the investment game. Left: using $c = 1$, $\gamma_1 = 1.4$, $\gamma_2 = 1.8$, $\rho_1 = 0.5$, $\rho_2 = 0.5$, $\rho_{11} = 1$, $\rho_{00} = 3$. Right: using $c = 0.5$, $\gamma_1 = 1.1$, $\gamma_2 = 1.4$, $\rho_1 = 0.4$, $\rho_2 = 0$, $\rho_{11} = 2$, $\rho_{00} = 0.4$

First note, alarmingly, how the two different approaches we might favour in computer simulations—time stepping or smoothing—give markedly different rates of change for z. Note also how the delayed or hysteretic implementations depend erratically on the relative size of delay/hysteresis across $x_1 = 0$ and $x_2 = 0$, i.e., the ratio ω.

The two different games in Fig. 4.5 exhibit markedly different behaviour. The behaviour outside $x_1 = x_2 = 0$ does not change significantly between these parameter values, remaining always pointing 'inwards' towards the origin. Despite this the speed of motion along $x_1 = x_2 = 0$ changes, both between the two parameter sets, and as we vary the implementation parameter $\varepsilon_2/\varepsilon_1 = \tan\frac{\omega\pi}{2}$. We could plot similar graphs with any of the parameters c, γ_i, ρ_i, on the horizontal axis in place of ω, and obtain similarly erratic curves.

The noisy implementations are missing from Fig. 4.5. The same two games are implemented with noisy delay or spatial noise (see Definition 4.1) in Fig. 4.6. Counterintuitively, the outcome is less erratic in the noisy conditions of Fig. 4.6 than in the regular conditions of Fig. 4.5. Noise appears to push the value of $\dot{z}$ towards the deterministic value obtained under smoothing. The significant random fluctuations are a result of the noise amplitude needing to be large to dampen the erratic variations in Fig. 4.5.

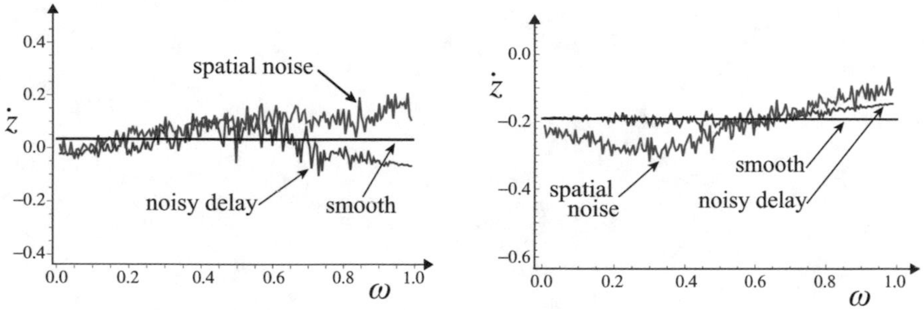

Fig. 4.6 Noisy implementations corresponding to Fig. 4.5

In Sects. 4.1 and 4.2 the implementation could push a system between two different possible behaviours. Here we see a wide range of erratically varying behaviours, and not only do different implementations select entirely different outcomes from within this set, but changing the parameters of the implementation itself, even slightly, selects vastly different behaviours. Different aspects of these rather striking results began emerging across a number of investigations [5, 6, 78, 84]. The erratic and sensitive dependence in the case of hysteresis was first studied in [6], and extended to other implementations in [84], contrasting with the more regular and predictable dynamics resulting from smoothing [5, 78].

The issues involved in the practical modelling of discontinuities clearly involve a number of factors. To begin addressing them we first need a qualitative way to characterize such behaviour, to which we turn now.

Chapter 5
Layers and Implementations

Consider a dynamical system modelled by an equation $\dot{\mathbf{x}} = \mathbf{f}(\mathbf{x})$, where the function $\mathbf{f}(\mathbf{x})$ is smooth except at some *discontinuity threshold* $\mathcal{D}$. Let the space of $\mathbf{x}$ be divided by $\mathcal{D}$ into open regions $\mathcal{R}_i$, where $i \in Z_N$ with Z_N denoting a set of N indices or *modes*. Let $\mathbf{f}(x)$ take a different functional form $\mathbf{f}^i(\mathbf{x})$ on each region $\mathcal{R}_i$, so

$$\dot{\mathbf{x}} = \mathbf{f}^i(\mathbf{x}) \quad \text{if} \quad \mathbf{x} \in \mathcal{R}_i, \quad i \in Z_N \,. \tag{5.1}$$

We assume that the functions $\mathbf{f}^i(\mathbf{x})$ are smooth in $\mathbf{x}$ over the closure $\overline{\mathcal{R}}_i$ of each region $\mathcal{R}_i$. A solution of such a system is depicted crossing a discontinuity threshold in Fig. 5.1(i), with (ii) showing its tangent vectors following the discontinuous vector field, and therefore changing direction abruptly at $\mathcal{D}$. Note that (5.1) does not yet define $\dot{\mathbf{x}}$ for $\mathbf{x} \in \mathcal{D}$.

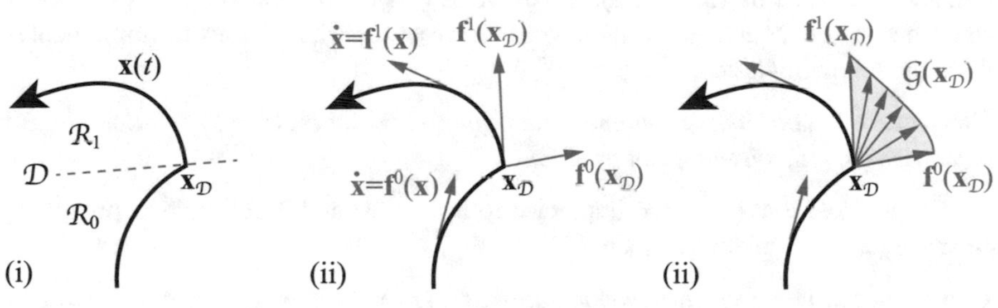

Fig. 5.1 (i) A trajectory crossing a discontinuity threshold $\mathcal{D}$, such that (ii) its tangent vectors are $\mathbf{f}^0$ and $\mathbf{f}^1$ either side of the discontinuity, and (iii) lie in a connected set $\mathcal{G}$ (their endpoints forming a curve connecting $\mathbf{f}^0$ to $\mathbf{f}^1$ as depicted) at a point $\mathbf{x}_\mathcal{D}$ on the discontinuity

To study how solutions of this system will behave at $\mathcal{D}$, we need to know *how* the tangent vector $\dot{\mathbf{x}}$ switches between the different vectors $\mathbf{f}^i(\mathbf{x})$ that it encounters near a point $\mathbf{x} \in \mathcal{D}$. We will do this by saying that $\dot{\mathbf{x}}$ sweeps continuously through some set of values as $\mathbf{x}$ crosses $\mathcal{D}$, as shown in Fig. 5.1(iii), and expressed as

M. R. Jeffrey, *Modeling with Nonsmooth Dynamics*,
Frontiers in Applied Dynamical Systems: Reviews and Tutorials 7,
https://doi.org/10.1007/978-3-030-35987-4_5

$$\dot{\mathbf{x}} \in \mathcal{G}(\mathbf{x}) \quad \text{where} \quad \mathcal{G}(\mathbf{x}) \supset \left\{ \mathbf{f}^i(\mathbf{x}) \;\text{ if }\; \mathbf{x} \in \overline{\mathcal{R}}_i,\; i \in Z_N \right\}, \tag{5.2}$$

where $\mathcal{G}(\mathbf{x})$ is a connected set that varies smoothly with $\mathbf{x}$.

This does not fully define $\mathcal{G}(\mathbf{x})$ itself, and we will look into specific ways of doing so in the following sections. In particular we will look at how different choices for $\mathcal{G}(\mathbf{x})$ relate to different modelling assumptions. All (5.2) tells us is that if $\mathbf{x}$ lies on the discontinuity threshold, then $\dot{\mathbf{x}}$ takes a set of values $\mathcal{G}(\mathbf{x})$ that interpolates in some way between any $\mathbf{f}^i(\mathbf{x})$ for which $\mathbf{x}$ lies on the boundary of $\mathcal{R}_i$, for any $i \in Z_N$. If $\mathbf{x}$ lies inside a region $\mathcal{R}_i$ and therefore outside the discontinuity threshold, then $\mathcal{G}(\mathbf{x})$ needs to contain only $\mathbf{f}^i(\mathbf{x})$, and it makes sense to define $\mathcal{G}(\mathbf{x})$ in such a way that it reduces to the right-hand side of (5.1) for $\mathbf{x} \notin \mathcal{D}$.

Before going on to define $\mathcal{G}(\mathbf{x})$ more closely, first let us extend (5.2) to describe our main topic of interest, that of perturbations of the idealized system.

Consider a system where a switch takes place over an ε-neighbourhood of the threshold $\mathcal{D}$, which we call the *switching layer* $\mathcal{D}^\varepsilon$. We can then refine Definition 4.1 as follows.

Definition 5.1. An **implementation** of the system (5.1) assigns a rule that defines (discrete or continuous) trajectories $\mathbf{x}(t)$ through any point $\mathbf{x}$ and satisfies

$$\dot{\mathbf{x}} \in \mathcal{G}^\varepsilon(\mathbf{x}) \qquad \text{such that} \qquad \mathcal{G}^\varepsilon(\mathbf{x}) = \bigcup_{\mathbf{u} \in B_\varepsilon(\mathbf{x})} \mathcal{G}(\mathbf{u}) \tag{5.3}$$

on a layer $\mathcal{D}^\varepsilon$ for some $\varepsilon > 0$, such that $\mathcal{D}^\varepsilon \supset \mathcal{D}$ and $\mathcal{D}^\varepsilon \to \mathcal{D}$ as $\varepsilon \to 0$, where $B_\varepsilon(\mathbf{x})$ denotes a ball of radius ε about $\mathbf{x}$.

An implementation may represent a more physically precise model of switching than the idealized system described by (5.1)–(5.2), or a method of simulating a model described by (5.1)–(5.2). In any case, one of the first goals of piecewise-smooth systems theory should then be to discover for what classes of implementations the following holds.

Conjecture 1. *Solutions of the implementation* (5.3) *lie ε-close to solutions of the piecewise-smooth system given by* (5.2).

We shall see that (1) is too imprecise to hold as stated, but the concepts above permit a more complete statement.

Conjecture 2. *For a given implementation, there exists a set $\mathcal{G}$ such that solutions of the implementation* (5.3) *lie ε-close to solutions of the piecewise-smooth system given by* (5.2).

The next few sections are essentially an exploration of these statements, obtaining insight into the classes of systems that (5.2) represents as a modelling paradigm. We proceed qualitatively. Rigorous proofs of the conjecture have been achieved so far only in such limited conditions that they have narrow applicability (but we will discuss these as we explore some of the implementations below). Alas a general proof of either conjecture would seem to be premature, most likely requiring a general representation of implementations that is far more explicit about their workings than Definition 4.1 or Definition 5.1, and relating broad classes of implementations to explicit forms for $\mathcal{G}^{\varepsilon}(\mathbf{x})$. Our modest aim here will instead be to gain insight into the dynamics of such implementations, with an eye to better formulating and proving these conjectures in the future.

We will also not prove that solutions to the dynamical systems above exist. A knowledge of the most general form of $\mathcal{G}$ or $\mathcal{G}^{\varepsilon}$ that guarantees the existence of solutions (beyond the convex and upper semi-continuous sets studied by Filippov in [51]) would help further reveal the kinds of system that can be modelled, but is beyond our ambitions here.

The concept of an implementation is related but not identical to the concept of a *regularization*. The purpose of regularization is, as the term suggests, to achieve regularity, in this case to make a discontinuous problem well-posed under certain limited conditions, usually with the aim of obtaining unique solutions. In analysis, for example, regularization is usually achieved with some kind of smoothing [31, 136], or in electronics usually with some kind of hysteresis or delay that ensures a finite interval of motion between each switching event [134, 148]. An implementation, on the other hand, merely seeks to render a system solvable without any expectation of uniqueness.

5.1 Linear Switching

A natural choice for the form of the set in (5.2) is to take the convex hull, which we call $\mathcal{F}$, of the vector fields across a discontinuity.

Definition 5.2. The **linear switching system** corresponding to (5.1) is given by

$$\dot{\mathbf{x}} \in \mathcal{F}(\mathbf{x}) \tag{5.4}$$
$$\text{where}\quad \mathcal{F}(\mathbf{x}) = \text{hull}\left\{ \mathbf{f}^i(\mathbf{x}) \ \forall\, i \ : \ \mathbf{x} \in \overline{\mathcal{R}}_i,\ i \in Z_N \right\},$$

where hull$\{Q\}$ denotes the convex hull of the set $\{Q\}$ (sometimes written as $\overline{\text{co}}\{Q\}$ or conv$\{Q\}$). Solutions of (5.4) will be called **linear solutions** of (5.1).

This is often referred to as a *Filippov system*, but for our purposes here we use the term *linear system* to highlight that we can express $\mathcal{F}$ as a linear function of switching multipliers μ_i, in the following way.

An explicit expression for the hull is given by a linear combination of the vector fields $\mathbf{f}^i(\mathbf{x})$,

$$\mathcal{F}(\mathbf{x}) = \sum_{i=1}^{N} \mu_i \mathbf{f}^i(\mathbf{x}) \quad \text{such that} \quad \sum_{i=1}^{N} \mu_i = 1\,, \tag{5.5a}$$

where

$$\mu_i = \begin{cases} 1 & \text{if } \mathbf{x} \in \mathcal{R}_i \,, \\ 0 & \text{if } \mathbf{x} \notin \overline{\mathcal{R}}_i \,, \\ [0,1] & \text{if } \mathbf{x} \in \delta\mathcal{R}_i \,, \end{cases} \tag{5.5b}$$

where $\delta\mathcal{R}_i = \overline{\mathcal{R}}_i/\mathcal{R}_i$ denotes the boundary of $\mathcal{R}_i$. So μ_i is a step function that is 1 on $\mathcal{R}_i$ and 0 outside, and set-valued on the boundary. Note that, because the μ_i's sum to unity, if $\mu_i = 1$ for some i, then $\mu_j = 0$ for all $j \neq i$. For the sake of defining the hull we let $\mu_i = [0,1]$ on $\delta\mathcal{R}_i$, but in Chap. 5 we will look at ways to assign specific values to $\mu_i \in [0,1]$ for each $i \in Z_N$, and hence assign specific values from the hull (5.5a) to $\dot{\mathbf{x}}$.

As we said for the discontinuous quantity ν following (1.15) in Sect. 1.1, we will not make the $\mathbf{x}$-dependence of μ_i explicit by writing $\mu_i(\mathbf{x})$, but must remember that it is a piecewise-constant for $\mathbf{x} \notin \mathcal{D}$ and is set-valued for $\mathbf{x} \in \mathcal{D}$ where, as the coefficients μ_i vary over the intervals [0, 1], the right-hand side of (5.5a) explores all values in the convex hull.

The hull is the smallest convex set that contains all of the $\mathbf{f}^i$s from (5.1). Extensive theory for such convex system was developed in [51], starting with the proof that solutions exist if $\mathcal{F}$ is non-empty, bounded, closed, convex, and upper semi-continuous. Moreover, making use of a set $\mathcal{F}_\delta$ that lies δ-close to $\mathcal{F}$, such that $\mathcal{F} \subset \mathcal{F}_\delta$ and $\mathcal{F}_0 = \mathcal{F}$, in [4, 51] it is proven that a solution exists through a point $\mathbf{x}_\delta$ that is δ-close to $\mathbf{x}$, and that the limit of a convergent sequence of solutions of (5.4) is itself a solution of (5.4). Most commonly the term *Filippov system* refers to a system where a point $\mathbf{x} \in \mathcal{D}$ can lie on the boundary of two regions $\mathcal{R}_1$ and $\mathcal{R}_2$ only, in which case the linear solutions of Definition 5.2 are mostly unique (except at certain singularities). They are no longer unique if $\mathbf{x} \in \mathcal{D}$ can lie on the boundary of more than two regions.

We refer the reader to [51] for the full theory of convex solutions rather than restating it here, because it is rather lengthy, and although a natural choice mathematically, the system (5.4) is actually a strong restriction of the possible systems allowed by (5.2). There is no physical or dynamical reason in general that restricts a system to follow a vector field lying in the convex hull $\mathcal{F}$. At the same time, in associating one switching multiplier μ_i with each vector field $\mathbf{f}^i$, the model (5.5a) contains somewhat more multipliers than are necessary to continuously transition between all of the modes $\mathbf{f}^i$, and instead it is possible to define sets that are both more general, and yet involve less unknown multipliers. This will matter when, in order to study solutions in detail, we come to fix the values of these multipliers by dynamical or physical considerations.

5.2 Nonlinear Switching

A common way that models like (5.1) arise in practice is when a function $\mathbf{F}$ depends on a number of discontinuous quantities. Let these quantities be a set of switching multipliers $\nu_1, \ldots, \nu_m$, defined without loss of generality as

$$\nu_j = \text{step}\,(\sigma_j(\mathbf{x})) \quad \text{for} \quad \sigma_j \neq 0\,, \tag{5.6}$$

recalling the definition (1.15), in terms of smooth scalar functions $\sigma_i(\mathbf{x})$ for $i = 1, \ldots, m$. We then write $\mathcal{G}(\mathbf{x})$ in (5.2) as

$$\mathcal{G}(\mathbf{x}) = \mathbf{F}(\mathbf{x}; \nu_1, \ldots, \nu_m)\,, \tag{5.7a}$$

where

$$\nu_j = \begin{cases} 1 & \text{if } \sigma_j(\mathbf{x}) > 0\,, \\ 0 & \text{if } \sigma_j(\mathbf{x}) < 0\,, \\ [0, 1] & \text{if } \sigma_j(\mathbf{x}) = 0\,. \end{cases} \tag{5.7b}$$

Similarly to the hull (5.5), for the sake of defining $\mathcal{G}(\mathbf{x})$ we let $\nu_j = [0, 1]$ on $\sigma_j = 0$, but in Chap. 5 we will look at ways to assign specific values to $\nu_j \in [0, 1]$ for each $j = 1, \ldots, m$, and hence assign specific values from the set (5.8) to $\dot{\mathbf{x}}$.

The right-hand side of (5.7a) is then set-valued on $\mathcal{D}$ due to the set-valuedness of (5.7b) at $\sigma_j = 0$, but taking a unique value if $\sigma_j \neq 0$ for all $j = 1, \ldots, m$. We may assume the σ_js and $\mathbf{F}$ to be defined for all $\mathbf{x}$ (or one may restrict these domains and the differentiability of these functions as a problem requires, but our analysis will mainly be local).

For (5.8) to correspond to (5.1) and (5.2) there must be a direct correspondence between each $\mathbf{f}^i(\mathbf{x})$ and each $\mathbf{F}(\mathbf{x}; \nu_1, \ldots, \nu_m)$ outside $\mathcal{D}$, that is when $\sigma_j \neq 0$ for all $j = 1, \ldots, m$. The simplest way this happens is if $\mathcal{G}(\mathbf{x})$ is a *convex canopy* of the fields $\mathbf{f}^i(\mathbf{x})$.

Definition 5.3. The set $\mathcal{G}(\mathbf{x})$ is a **convex canopy** of the vector fields $\mathbf{f}^i(\mathbf{x})$ indexed by $i \in Z_N$ in (5.1) if it is expressible in the form (5.7a) in terms of switching multipliers $\nu_1, \ldots, \nu_m$, given by (5.7b), such that $N = 2^m$, and such that: $\mathbf{F}$ has continuous dependence on $\mathbf{x}$ and on the ν_js, and there is a one-to-one correspondence between each of the N functions $\mathbf{f}^i$ with $i \in Z_N$, and the 2^m vector fields $\mathbf{F}(\mathbf{x}; \nu_1, \ldots, \nu_m)$ in which each ν_j takes value 0 or 1.

We then write the following.

Definition 5.4. The **nonlinear switching system** corresponding to (5.1) is given by

$$\dot{\mathbf{x}} \in \mathcal{G}(\mathbf{x}) \quad \text{where} \quad \mathcal{G}(\mathbf{x}) = \mathbf{F}(\mathbf{x}; \nu_1, \ldots, \nu_m)\,, \tag{5.8}$$

where $\mathbf{F}$ depends smoothly on $\mathbf{x}$ and on switching multipliers ν_j as defined in (5.7b). Solutions of (5.8) will be called **nonlinear solutions** of (5.1).

The correspondence required in Definition 5.3 can be achieved by associating each $i \in Z_N$ with a binary string $i_1 \ldots i_m$ of 0s and 1s, by either

$$(a)\quad i = 1 + \sum_{j=1}^{m} 2^{j-1} i_j \qquad \text{or} \qquad (b)\quad i = i_1 \dots i_m\,, \tag{5.9}$$

where (a) simply numbers regions so $i \in Z_N = \{1, 2, \dots, N\}$, while (b) identifies i as the binary string $i = i_1 i_2 \dots i_m$ itself.

The simplest example of a canopy is then given by writing

$$\mathcal{G}(\mathbf{x}) = \sum_{i=1}^{2^m} \mu_i \mathbf{f}^i(\mathbf{x}) \quad \text{where} \quad \mu_i = \prod_{j=1}^{m} \nu_j^{i_j}(1-\nu_j)^{1-i_j}\,, \tag{5.10}$$

with i and $i_1 \dots i_m$ related by either rule in (5.9). Thus when all σ_js are non-vanishing, $\mathcal{G}(\mathbf{x})$ has a unique value $\mathbf{f}^i(\mathbf{x})$. If any σ_j vanishes, then the corresponding multiplier ν_j is set-valued, each μ_i either vanishes or is set-valued, and hence $\mathcal{G}$ is set-valued.

The canopy given by (5.10) is typically a subset of the hull in (5.5a), that is $\mathcal{G}(\mathbf{x}) \subseteq \mathcal{F}(\mathbf{x})$. In fact the canopy is typically a lower dimensional set, as rather than $N-1$ coefficients μ_i in the hull (5.5a), we now have only $m-1$ coefficients ν_j, with $N = 2^m$. Canopies still given by (5.7a) but taking values outside the hull (5.5a) are also possible if $\mathbf{F}$ includes *hidden* terms, which we will come to in Sect. 5.3.

Although the expression (5.10) may look somewhat opaque at first, it is easy to expand for any given m and the resulting expressions are rather easier to understand. For convenience we give the cases $m = 1$ in the different indexing systems in Appendix B.

This system now consists of m discontinuity thresholds $\mathcal{D}_j$ that comprise the discontinuity surface $\mathcal{D}$,

$$\begin{aligned} \mathcal{D}_j &= \left\{\mathbf{x} \in \mathbb{R}^n \;:\; \sigma_j(\mathbf{x}) = 0\right\} \qquad \text{and} \\ \mathcal{D} &= \{\mathbf{x} \in \mathbb{R}^n \;:\; \sigma_1(\mathbf{x}) \dots \sigma_m(\mathbf{x}) = 0\}\,, \end{aligned} \tag{5.11}$$

and clearly this assumes that the discontinuity surface is a manifold expressible as a union of submanifolds $\mathcal{D} = \mathcal{D}_1 \cup \dots \cup \mathcal{D}_m$.

The convex *canopy* of the vector fields $\mathbf{f}^i$ is a natural choice for a set interpolating between the fields $\mathbf{f}^i$, as just the multi-linear interpolation between the $\mathbf{f}^i$'s in terms of the coefficients ν_j, in some ways more natural than the convex hull (5.5a) which is strictly linear in it coefficients μ_i but requires a larger number of them. As such the expression (5.10) result has been arrived at independently from different viewpoints by several authors, perhaps first in [5] as a way of 'blending' the fields $\mathbf{f}^i$ to seek unique motion along $\mathcal{D}$, as well as in [37, 38] as a way of facilitating computation, in [78, 81] as a study of nonlinearity in switching, and in [98, 141] to derive equivalence classes of regularized systems.

Functions like $\mathbf{F}(\mathbf{x}; \nu_1, \dots, \nu_m)$ appear naturally in many applications, wherever a physical parameter ν_j jumps in value as some scalar quantity σ_j crosses a threshold. The 'three experiments' in Chap. 4 took this form, and in electronic control this is familiar as Utkin's formulation of variable structure systems (see, e.g., [127, 148]).

5.3 Hidden Terms: The 'Ghosts' of Switching

The expression (5.10) for $\mathcal{G}$ is allowed to have nonlinear dependence on the switching multipliers $\nu_j = \text{step}(\sigma_j)$. Let us now ask what this nonlinearity signifies, and how this relates to the problem of defining the value of step(0).

Consider that we have a vector field that behaves as $\dot{\mathbf{x}} = \mathbf{F}(\mathbf{x}; \text{step}(\sigma(\mathbf{x})))$ for $\sigma \neq 0$, and a modelling parameter that behaves as $\nu = \text{step}(\sigma(\mathbf{x}))$. Can we simply assume that an adequate model is given by $\dot{\mathbf{x}} = \mathbf{F}(\mathbf{x}; \nu)$? What is the difference if we instead model this as $\dot{\mathbf{x}} = \mathbf{F}(\mathbf{x}; \nu^p)$ for some $p \in \mathbb{N}$, since we can also write $\nu^p = \text{step}(\sigma(\mathbf{x}))$? This will have non-trivial consequences on the discontinuity threshold, where ν varies over the interval $[0, 1]$.

The difference between any monomial ν^p and the linear term ν can be written as

$$\nu^p - \nu = \nu(\nu - 1) \sum_{r=0}^{p-2} \nu^r \qquad \text{for } p \geq 2 . \tag{5.12}$$

This is what we called in Sect. 1.1 a *hidden* term, since it vanishes for all $\sigma \neq 0$, as highlighted by the factorization on the right-hand side where either ν or $\nu - 1$ vanishes if ν is 0 or 1. The hidden term need not vanish inside the switching layer, however, where $0 \leq \nu \leq 1$. Let us now write this more formally.

Definition 5.5. A **hidden** term $\mathbf{H}(\mathbf{x}; \nu)$ associated with a discontinuity threshold $\mathcal{D}$ vanishes everywhere outside $\mathcal{D}$. If $\mathbf{x} \in \mathcal{D}$ at a point where $\mathcal{D}$ is an intersection $\mathcal{D}_1 \cap \cdots \cap \mathcal{D}_m$ of manifolds $\mathcal{D}_j = \left\{\mathbf{x} \in \mathbb{R}^n : \sigma_j(\mathbf{x}) = 0\right\}$, then we can write this as

$$\mathbf{H}(\mathbf{x}; \nu)\sigma_1(\mathbf{x}) \ldots \sigma_m(\mathbf{x}) = 0 \quad \text{for any} \quad \mathbf{x} . \tag{5.13}$$

That is, either some $\sigma_j = 0$ and $\mathbf{x}$ lies on the discontinuity threshold, or $\sigma_j \neq 0$ for all j in which case $\mathbf{H}(\mathbf{x}; \nu) = 0$.

Using hidden terms we can distinguish between different systems such as $\dot{\mathbf{x}} = \mathbf{F}(\mathbf{x}; \nu)$ and $\dot{\mathbf{x}} = \mathbf{F}(\mathbf{x}; h(\nu))$, where h is any function of ν that behaves like $h(\nu) = \text{step}(\sigma(\mathbf{x}))$, for example, $h(\nu) = \nu^p$ for $p \in \mathbb{N}$.

More generally we have the following.

Lemma 5.3. *If a system* $\dot{\mathbf{x}} = \mathbf{F}(\mathbf{x}; \nu)$ *can be expressed in terms of a multiplier* $\nu = \text{step}(\sigma(\mathbf{x}))$ *for some smooth function* σ*, such that* $\mathbf{F}$ *is k-times differentiable with respect to* ν*, then we can decompose it into the convex combination of the fields* $\mathbf{f}^0(\mathbf{x}) \equiv \mathbf{F}(\mathbf{x}; 0)$ *and* $\mathbf{f}^1(\mathbf{x}) \equiv \mathbf{F}(\mathbf{x}; 1)$*, plus a k-times differentiable hidden term* $\mathbf{H}(\mathbf{x}; \nu)$*, as*

$$\dot{\mathbf{x}} = \mathbf{F}(\mathbf{x}; \nu) = \nu\mathbf{f}^1(\mathbf{x}) + (1 - \nu)\mathbf{f}^0(\mathbf{x}) + \mathbf{H}(\mathbf{x}; \nu) . \tag{5.14}$$

Proof. The result is straightforward from Definition 5.5. It is clear from (5.14) that $\mathbf{F}(\mathbf{x}; 0) \equiv \mathbf{f}^0(\mathbf{x})$ and $\mathbf{F}(\mathbf{x}; 1) \equiv \mathbf{f}^1(\mathbf{x})$, and therefore that $\mathbf{H}(\mathbf{x}; 0)$ and $\mathbf{H}(\mathbf{x}; 1)$ must vanish, so $\mathbf{H}$ is a hidden term. The differentiability of $\mathbf{H}$ with respect to ν follows directly from that of $\mathbf{F}$. □

More useful than the result itself is to use this to derive an expression for $\mathbf{H}$. If $\mathbf{F}$ is k-times differentiable with respect to ν, then we can expand it as a polynomial

$$\dot{\mathbf{x}} = \mathbf{F}(\mathbf{x};\nu) = \sum_{r=0}^{k} \mathbf{c}_r(\mathbf{x})\nu^r , \tag{5.15}$$

for some vector fields $\mathbf{c}_r(\mathbf{x})$. We can express the first two of these, $\mathbf{c}_0$ and $\mathbf{c}_1$, in terms of the vector fields $\mathbf{f}^0$ and $\mathbf{f}^1$ either side of the discontinuity by evaluating $\mathbf{F}$ at $\nu = 0$ and $\nu = 1$. Using (5.15) this gives $\mathbf{f}^0(\mathbf{x}) = \mathbf{F}(\mathbf{x};0) = \mathbf{c}_0$ and $\mathbf{f}^1(\mathbf{x}) = \mathbf{F}(\mathbf{x};1) = \sum_{r=0}^{k}\mathbf{c}_r$, which re-arranges to $\mathbf{c}_0 = \mathbf{f}^0$ and $\mathbf{c}_1 = \mathbf{f}^1 - \mathbf{f}^0 - \sum_{p=2}^{k}\mathbf{c}_p$. Substituting these into (5.15) we obtain (5.14), where

$$\mathbf{H}(\mathbf{x};\nu) = \sum_{p=2}^{k} \mathbf{c}_p(\mathbf{x})(\nu^p - \nu) = \nu(\nu-1)\sum_{p=2}^{k}\sum_{r=0}^{p-2} \mathbf{c}_p(\mathbf{x})\nu^r , \tag{5.16}$$

using the identity (5.12) to extract the factor $\nu(\nu-1)$ in $\mathbf{H}$. Thus $\mathbf{H}$ is a hidden term since $\mathbf{H}(\mathbf{x};0) = \mathbf{H}(\mathbf{x};1) = 0$, and hence $\mathbf{H}(\mathbf{x};\nu)\sigma(\mathbf{x}) = 0$ for all $\mathbf{x}$.

Nonlinear dependence on discontinuous multipliers therefore gives us a way to characterize different systems that *appear* the same almost everywhere, but differ at a discontinuity threshold, distinguished via hidden terms. The result generalizes easily to several switching multipliers ν_j simply by forming a multi-variable polynomial expansion in powers of ν_j.

The hidden term (5.16) has a slightly different formula to that we obtained in (1.11)–(1.12) in Sect. 1.1, but they are consistent. We developed (1.11) as an infinite series for a scalar problem $\dot{x} = F(x)$ with switching threshold $x = 0$, so to see the equivalence we need to merely make the scalars x, F, a_n, b_n, into vectors $\mathbf{x}, \mathbf{F}, \mathbf{a}_n, \mathbf{b}_n$, let the threshold be at $\sigma(\mathbf{x}) = 0$, and truncate the series at $O\left(\nu^k\right)$. Clearly then the linear part of the series in (1.11) and (5.14) is equivalent with $\mathbf{f}^1(\mathbf{x}) = \mathbf{a}_0(\mathbf{x})$ and $\mathbf{f}^0(\mathbf{x}) = \mathbf{b}_0(\mathbf{x})$. The hidden terms are both of the form

$$\mathbf{H}(\mathbf{x};\nu) = \nu(1-\nu)\mathbf{G}(\mathbf{x};\nu) \quad \text{where} \quad \mathbf{G}(\mathbf{x};\nu) = \sum_{r=0}^{k-2} \mathbf{d}_r(\mathbf{x}) , \tag{5.17}$$

with coefficients $\mathbf{d}_r(\mathbf{x})$ related to $\mathbf{c}_r$ in (5.16) and $\mathbf{a}_r(\mathbf{x})$, $\mathbf{b}_r(\mathbf{x})$, in (1.12), by

$$\mathbf{d}_r(\mathbf{x}) = -\sum_{p=r+2}^{k} \mathbf{c}_p(\mathbf{x}) = \mathbf{b}_{r+1}(\mathbf{x}) + (-1)^r \sum_{p=r}^{k-2} \tfrac{p!}{r!(p-r)!}\mathbf{a}_{p+1}(\mathbf{x}) . \tag{5.18}$$

Thus we are beginning to see how different modelling approaches lead to the same fundamental form for the expansion of a piecewise-smooth function (5.14), with a hidden term $\mathbf{H} = \nu(1-\nu)\mathbf{G}$, and it is only the function $\mathbf{G}$ we obtain that changes.

We will see a few more forms for **G** yet in Sect. 6.3, obtained by considering implementations that smooth the discontinuity in different ways.

Such in-depth analysis of dynamics at the discontinuity becomes particularly necessary if solutions dwell upon the discontinuity threshold for significant intervals of time, as seen in all of the examples in Chap. 4. That motion is known as *sliding* along the discontinuity, and it is one of the oldest and most important notions in nonsmooth dynamics. To deal with the various behaviours in Chap. 4 associated with nonlinearity of the vector field or with the implementation of the discontinuity, we need to significantly expand that standard concept of sliding. We shall do this in Chap. 6, before applying these ideas specifically to the prototypes from Chap. 4.

Chapter 6
Ideal and Non-ideal Sliding

Sliding—motion along the discontinuity threshold—is central to the most novel phenomena of nonsmooth dynamics. The theory was largely developed in Filippov's work [51], but seems to originate in earlier Russian texts, perhaps first from G. N. Nikol'skii [109] (see also discussion in [4, 107, 148]). The standard definition would describe sliding as motion along an ideal threshold $\mathcal{D}$. We shall define it as follows.

Definition 6.1. A solution of an implementation (5.3) of a piecewise-smooth system (5.1) is said to **slide** if it evolves inside the layer $\mathcal{D}^\varepsilon$ for a time $\Delta t = O(1)$ (i.e., a time not vanishing as $\varepsilon \to 0$).

This allows us to discuss sliding in implementations as well as in the ideal discontinuous system. Intuitively, sliding occurs because solutions tend towards some invariant set in the layer $\mathcal{D}^\varepsilon$ around the discontinuity threshold $\mathcal{D}$. More precisely we can state the following.

Lemma 6.1. *Consider system* (5.1) *on an open region W, defined piecewise on regions $\mathcal{R}_i = \{\mathbf{x} \in \mathbb{R}^n : i = \text{step}(\sigma(\mathbf{x})),\ \sigma \neq 0\}$ in terms of vector fields $\mathbf{f}^i$ that are differentiable on $\sigma \geq -\varepsilon$ (for $i = 1$) and $\sigma \leq +\varepsilon$ (for $i = 0$), and a scalar function σ differentiable for all* $\mathbf{x}$. *Take an implementation of* (5.1) *on W according to Definition 5.1. If the discontinuity threshold $\mathcal{D} = \{\mathbf{x} : \sigma = 0\}$ is either attracting or repelling with respect to the vector fields $\mathbf{f}^i$ for $\sigma \to 0$ on W, and the vector fields $\mathbf{f}^i$ have non-vanishing components normal to $\mathcal{D}$ for $\sigma \to 0$ on W, then there exists some $E > 0$ such that the layer $\mathcal{D}^\varepsilon$, where $\sigma = O(\varepsilon)$, is invariant inside W for $0 < \varepsilon < E$.*

Proof. The attractivity of $\mathcal{D}$ implies $\mathbf{f}^1(\mathbf{x}) \cdot \nabla\sigma < 0 < \mathbf{f}^0(\mathbf{x}) \cdot \nabla\sigma$, and repulsivity of $\mathcal{D}$ implies the opposite signs, given that the normal components $\mathbf{f}^i \cdot \nabla\sigma$ are non-vanishing, evaluated at $\mathbf{x}$ as $\sigma \to 0$. Since $\mathbf{f}^0(x)$ and $\mathbf{f}^1(x)$ are differentiable, at any point $\mathbf{x} \in W$ on $\sigma = 0$, there exists $\tilde{E}(\mathbf{x}) > 0$ such that $\mathbf{f}^1(\mathbf{u}) \cdot \nabla\sigma < 0 < \mathbf{f}^0(\mathbf{u}) \cdot \nabla\sigma$ for all $\mathbf{u}$ such that $|\mathbf{u} \cdot \nabla\sigma| < \tilde{E}(\mathbf{x})$. Let $\tilde{E}(\mathbf{x})$ be the largest such value at each $\mathbf{x}$, and

M. R. Jeffrey, *Modeling with Nonsmooth Dynamics*,
Frontiers in Applied Dynamical Systems: Reviews and Tutorials 7,
https://doi.org/10.1007/978-3-030-35987-4_6

let E be the infimum of all $\tilde{E}(\mathbf{x})$ for $\mathbf{x} \in \mathcal{D} \cap W$. Then for any $\varepsilon < E$ the vector fields satisfy $\mathbf{f}^1(\mathbf{x}) \cdot \nabla\sigma < 0$ for $\mathbf{x}$ such that $\sigma < +\varepsilon$, and $\mathbf{f}^0(\mathbf{x}) \cdot \nabla\sigma > 0$ for $\mathbf{x}$ such that $\sigma > -\varepsilon$, therefore the region $|\sigma(\mathbf{x})| < \varepsilon$ is invariant. □

This suggests that if we can define a switching multiplier (some μ or ν), and a dynamics on it in the switching layer, then sliding constitutes an invariant of the dynamics confined to the layer, when the dynamics would otherwise carry trajectories through the layer. Even if we cannot define the layer dynamics in closed form, we can still consider the attracting or repelling objects it forms that constitute sliding.

Lemma 6.1 treats only the case when $\mathcal{D}$ is attracting or repelling with respect to the flows outside it, but attractivity/repulsivity is neither necessary nor sufficient for sliding to take place. The extension of Lemma 6.1 to a discontinuity threshold $\mathcal{D}$ formed by the intersection of manifolds $\mathcal{D}_1 \cap \cdots \cap \mathcal{D}_m$, for example, is quite straightforward, this also has been mainly considered only under conditions of uniformly attraction of $\mathcal{D}$ with respect to the surrounding flows, e.g., in [5, 38].

The situation in general is greatly more complicated. The discontinuity threshold $\mathcal{D}$ need not be attracting or repelling for sliding to occur (as we saw in the example of 'sticky genes' in Sect. 1.3, see also [81]). Moreover, an intersection of manifolds $\mathcal{D}_1 \cap \cdots \cap \mathcal{D}_m$ can be attracting without the individual thresholds $\mathcal{D}_j$ being attracting, for example, if the flow spirals around a point $x_1 = x_2 = 0$ by crossing through discontinuity thresholds $x_1 = 0$ and $x_2 = 0$ (see, e.g., [36, 51]). The permutations are enormous and no substantial accounting of the possibilities has been made. Perhaps the most ambitious steps in this direction are in [66], where the authors classify the behaviours a solution can exhibit once it enters an intersection in a planar system.

The only general statement that can be made is that if the set $\mathcal{G}$ from (5.2) has an intersection with the tangent space $\mathcal{T}_\mathcal{D}$ of the discontinuity threshold at a given point $\mathbf{x} \in \mathcal{D}$,

$$\mathcal{G}(\mathbf{x}) \cap \mathcal{T}_\mathcal{D}(\mathbf{x}) \neq \emptyset \,, \tag{6.1}$$

then sliding motion is possible in (5.1). Similarly if the set $\mathcal{G}^\varepsilon$ of a given implementation from (5.3) has an intersection with the tangent space $\mathcal{T}_\mathcal{D}$ of the discontinuity threshold at a given point $\mathbf{x} \in \mathcal{D}$,

$$\mathcal{G}^\varepsilon(\mathbf{x}) \cap \mathcal{T}_\mathcal{D}(\mathbf{x}) \neq \emptyset \,, \tag{6.2}$$

then sliding motion is possible in the implementation of (5.1). A solution arriving at $\mathcal{D}^\varepsilon$ evolves onto some attractor that lies in $\mathcal{D}^\varepsilon$, on which solutions $\mathbf{x}(t)$ evolve in a direction ε-close to the tangent space to $\mathcal{D}$. Thus we formalize the notion of sliding more precisely as follows.

Definition 6.2. A solution $\mathbf{x}(t)$ of (5.1) is said to **slide** along the discontinuity threshold $\mathcal{D}$ if $\dot{\mathbf{x}}(t) \in \mathcal{G}(\mathbf{x}) \cap \mathcal{T}_\mathcal{D}(\mathbf{x})$. A solution $\mathbf{x}(t)$ of an implementation of (5.1) is said to **slide** along the discontinuity threshold $\mathcal{D}$ (more strictly along the layer $\mathcal{D}^\varepsilon$ approximating $\mathcal{D}$) if $\dot{\mathbf{x}}(t) \in \mathcal{G}^\varepsilon(\mathbf{x}) \cap \mathcal{T}_\mathcal{D}(\mathbf{x})$.

By saying $\dot{\mathbf{x}}(t) \in \mathcal{G}^\varepsilon(\mathbf{x}) \cap \mathcal{T}_\mathcal{D}(\mathbf{x})$, we mean that the tangent vector along a trajectory, $\dot{\mathbf{x}}(t)$, which must lie in $\mathcal{G}^\varepsilon(\mathbf{x})$, lies tangent to the discontinuity threshold $\mathcal{D}$, implying that $\mathbf{x}(t)$ evolves along $\mathcal{D}^\varepsilon$ as t changes.

To find out whether sliding *will* occur in either situation requires a look at the dynamics, to find out whether the vector fields in the sets (6.1) and (6.2) possess attractors or repellers in the layer that can be followed by any solutions $\mathbf{x}(t)$. We explore the different approaches to this over Sects. 6.1 and 6.3.

The experiments in Chap. 4 show sliding under the implementations from Definition 4.1. In each case sliding occurs in a layer $\mathcal{D}_j^{\varepsilon_j}$ that forms an order ε_j-neighbourhood around an ideal discontinuity threshold $\mathcal{D}_j = \{\mathbf{x} : \sigma_j(\mathbf{x}) = 0\}$. Regardless of implementation, the attractivity of the layer implies the existence of local attractors inside it. The effects of Lemma 6.1 are seen in Figs. 4.1 to 7.4, with sliding occurring along the threshold $x = 0$ in Fig. 4.1, the thresholds $x_1 = 0 < x_2$ and $x_2 = 0 < x_1$ in Figs. 4.2 and 4.3, and the threshold $x_1 = 0 < x_2$ in Figs. 4.4, 4.5, and 4.6. With smoothing, these attractors are normally hyperbolic manifolds or equilibria. With hysteresis, the attractors are cycles oscillating between the boundaries $x_1 = \pm\varepsilon_1$ and $x_2 = \pm\varepsilon_2$, therefore reducible to return maps on those surfaces. For time stepping or delay the attractors can be bounded within definite ε_1 and ε_2 neighbourhoods of $x_1 = 0$ and $x_2 = 0$, respectively, and are described by piecewise-smooth two-dimensional maps on the plane.

Before returning to look at these experiments again closely, we need to build up a more general picture of the dynamics inside the switching layer that result from these definitions of layers, implementations, and sliding.

6.1 Sliding Perspective I: The Piecewise-Smooth System

Given the description of a piecewise-smooth system in terms of switching multipliers, as given by (5.2) with (5.8), we can analyse dynamics at the discontinuity as follows.

In (1.15) we defined the multipliers $\nu_j = \text{step}(\sigma_j)$ only as taking values $\nu_j \in [0, 1]$ at $\sigma_j = 0$. We shall now ask how each ν_j varies across the interval $[0, 1]$ and derive dynamics on them induced by the vector fields in $\sigma_j \neq 0$. (Clearly outside the discontinuity surface the ν_js are piecewise-constant and so obey $\dot{\nu}_j = 0$ for any j).

On a discontinuity threshold $\mathcal{D}_j$, we can treat each ν_j as a 'blow-up' variable of the discontinuity set $\sigma_j = 0$. This is done by letting $\sigma_j = \varepsilon_j \nu_j$ for some small $\varepsilon_j \geq 0$, so that the discontinuity occurs across an interval $\sigma_j \in [0, \varepsilon_j] \to 0$ as $\varepsilon_j \to 0$. (This method is developed in [79, 81] but is essentially just a scaling of the quantity σ_j that maps its values on $\sigma_j \in [0, \varepsilon] \to 0$ to values $\nu_j \in [0, 1]$, and has no doubt been used earlier, e.g., in [148]. The term 'blow-up' itself appears to originate from singular perturbation literature [40]). We then use this to find the dynamics of ν_j on the interval $\nu_j \in [0, 1]$.

At a point where $\mathcal{D}$ is a codimension m manifold, let $\mathbf{x}|_{\mathcal{D}} \in \mathbb{R}^{n-m}$ denote the space of $\mathbf{x}$ restricted to $\mathcal{D}$. The multipliers ν_j in (5.10) lie on intervals $[0, 1]$, so the dynamics on $\mathcal{D}$ can be said to take place inside a *switching layer*

$$\mathcal{D}^{\varepsilon} = \{ (\nu_1, \ldots, \nu_m, \mathbf{x}|_{\mathcal{D}}) \in [0,1]^m \times \mathbb{R}^{n-m} \} . \tag{6.3}$$

For economy of nomenclature we use the term 'switching layer' to describe both the parameterization of $\mathcal{D}$ given by (6.3) in the piecewise-smooth system, and the region around $\mathcal{D}$ given by (5.3) in the implementation, each with an associated small parameter ε. The concepts are closely related, and one may refer to the 'switching layer of the piecewise-smooth system' or the 'switching layer of the implementation' if necessary to avoid confusion.

At a point on a discontinuity threshold where $\sigma_1 = \cdots = \sigma_m = 0$ for some $m \geq 1$, let us take local coordinates $\mathbf{x} = (X, \underline{x})$, where $X = (\sigma_1, \ldots, \sigma_m)$ and $\underline{x} \in \mathbb{R}^{n-m}$, so $\mathbf{x}_{\mathcal{D}} = (0, \ldots, 0, \underline{x})$. We then obtain on $\mathcal{D}$ a switching layer

$$\mathcal{D}^{\varepsilon} = \left\{ (\nu_1, \ldots, \nu_m, \underline{x}) \in [0,1]^m \times \mathbb{R}^{n-m} \right\} , \tag{6.4}$$

with each multiplier ν_j constituting the blow-up variable of the set $\sigma_j = 0$, as some small parameter $\varepsilon_j \to 0^+$. The switching layer is n dimensional, and differentiating $\sigma_j = \varepsilon_j \nu_j$ according to $\dot{\mathbf{x}} = \mathbf{F}(\mathbf{x}; \nu_1, \ldots, \nu_m)$ in these coordinates, given $\dot{\sigma}_j = \mathbf{F} \cdot \nabla \sigma_j = F_j$, we have

$$\begin{aligned} \varepsilon_j \dot{\nu}_j = \dot{\sigma}_j &= F_j(\varepsilon_1 \nu_1, \ldots, \varepsilon_m \nu_m, \underline{x}; \nu_1 \ldots, \nu_m) \\ &= F_j(0, \ldots, 0, \underline{x}; \nu_1 \ldots, \nu_m) + O(\varepsilon_1, \ldots, \varepsilon_m) \ . \end{aligned} \tag{6.5}$$

Neglecting the higher order term for $\varepsilon_j \to 0$, we obtain a well-defined layer system on $X = (0, \ldots, 0)$,

$$\varepsilon_j \dot{\nu}_j = F_j(0, \ldots, 0, \underline{x}; \nu_1, \ldots, \nu_m) \ , \quad j = 1, \ldots, m, \tag{6.6a}$$

$$\dot{\underline{x}} = \underline{F}(0, \ldots, 0, \underline{x}; \nu_1, \ldots, \nu_m) \ , \tag{6.6b}$$

up to terms of order $\varepsilon_1, \ldots, \varepsilon_m$, on the right-hand side. We recall that in this notation $\mathbf{F} = (F, \underline{F})$ and $F = (F_1, \ldots, F_m) = (\mathbf{F} \cdot \nabla \sigma_1, \ldots, \mathbf{F} \cdot \nabla \sigma_m)$.

In deriving this system we have fixed a very simple relationship between ν_j and σ_j, namely a linear (if singular as $\varepsilon_j \to 0$) mapping. We can do this without loss of generality because, through hidden terms, we are able to express any more complex functional relationship between some switching multiplier ν_j and the switching function σ_j using nonlinearity.

Say, for example, a vector field has a component $F_j = 1 + \mathrm{step}(\sigma_j)$, representing perhaps the reaction force from an object stuck to a surface with $\sigma_j = 0$, and say that F_j is known to pass through zero twice as the function σ_j changes sign. Clearly the function $F_j = 1 + \nu_j$ does not satisfy this, as $\nu_j \in [0, 1]$ implies $1 + \nu_j \in [1, 2]$. The function $F_j = 1 + \nu_j - r\nu_j(1 - \nu_j)$, however, varies over $F_j \in [\frac{1}{4}(6 - r^{-1} - r), 2]$ for $r > 1$, and if, say, F_j is known to vanish at some $\nu_j = k$, then $r = (1 + k)/(1 - k)k$ provides this.

With the dynamics at a discontinuity threshold $\mathcal{D}$ thus described by (6.6), sliding occurs if there exist fix points of (6.6a). These are sets of points satisfying $\dot{\nu}_j = 0$, and they generate *sliding manifolds*

$$\mathcal{M} = \left\{ \begin{array}{l} (\underline{x}, \nu_1, \ldots, \nu_m) \in \mathbb{R}^{n-m} \times [0,1]^m \ \text{ such that} \\ F_j(0, \ldots, 0, \underline{x}; \nu_1, \ldots, \nu_m) = 0, \ \ j = 1, \ldots, m \end{array} \right\} , \tag{6.7}$$

which are invariant wherever they are normally hyperbolic (see [81]). (Recall that $\mathbf{x}|_{\mathcal{D}} = (0, \ldots, 0, \underline{x})$ denotes $\mathbf{x}$ restricted to $\mathcal{D}$, and $\mathbf{F} \cdot \nabla\sigma_j = F_j$).

On $\mathcal{M}$ the dynamics takes the form of a *sliding mode*, given by

$$0 = F_j(0, \ldots, 0, \underline{x}; \nu_1, \ldots, \nu_m) , \quad j = 1, \ldots, m, \tag{6.8a}$$
$$\dot{\underline{x}} = \underline{F}(0, \ldots, 0, \underline{x}; \nu_1, \ldots, \nu_m) , \tag{6.8b}$$

Because (6.7) consists of m equations $F_1 = \cdots = F_m = 0$, in m unknowns given by the switching multipliers $\nu_1, \ldots, \nu_m$, they typically define a well-defined set $\mathcal{M}$ inside a given switching layer. This set may consist of branches of different stability (determined by considering the eigenvalues of the matrix $\partial(F_1, \ldots, F_m)/\partial(\nu_1, \ldots, \nu_m)$), connected by non-hyperbolic points (where those eigenvalues have zero real part).

In a system of many switches $\nu_j = \text{step}(\sigma_j)$, a solution may evolve between places where the discontinuity threshold $\mathcal{D}$ consists of an intersection of m different submanifolds $\mathcal{D}_j$, as defined in (5.11). On each different such region of $\mathcal{D}$ for different m, we first blow up each $\mathcal{D}_j$ into a switching layer, then derive the sliding modes, which occupy local sliding manifolds $\mathcal{M}$ of dimension $\mathbb{R}^{n-m}$.

If there is nonlinear dependence on the multipliers ν_j, then there may exist multiple equilibria, periodic or complex attractors, undergoing bifurcations and any other nonlinear phenomena inside the switching layer.

We will apply these ideas later when we look at some examples of linear versus nonlinear dynamics on the discontinuity threshold.

6.2 Sliding Perspective II: Hybrid Implementations

We define an implementation as *hybrid* if it cannot be expressed by means of a set of ordinary differential equations alone, but instead is given by a hybrid of the system

$$\dot{\mathbf{x}} = \mathbf{f}^i(\mathbf{x}) \quad \text{if} \quad \mathbf{x} \in \mathcal{R}_i^{\varepsilon} , \quad i \in Z_N , \tag{6.9}$$

along with a map

$$\dot{\mathbf{x}} = \mathbf{f}^i(\mathbf{x}) \quad \text{with} \quad i \mapsto \begin{cases} i & \text{if event}(\mathbf{x}; \varepsilon) = \text{false} , \\ \Psi(\mathbf{x}; i) & \text{if event}(\mathbf{x}; \varepsilon) = \text{true} , \end{cases} \tag{6.10}$$

with Z_N as before being some discrete set of N labels, and with the regions $\mathcal{R}_i^\varepsilon$ obeying $\mathcal{R}_i^\varepsilon \supset \mathcal{R}_i$, such that the switching layer $\mathcal{D}^\varepsilon$ is formed by the overlap of two or more regions $\mathcal{R}_i^\varepsilon$, though $\mathcal{R}_i^0 = \mathcal{R}_i$.

The system evolves as $\dot{\mathbf{x}} = \mathbf{f}^i(\mathbf{x})$ until a condition 'event$(\mathbf{x};\varepsilon)$' is satisfied, then i is updated to a new mode $\Psi(\mathbf{x};i)$. The hysteretic, delay, stochastic, and time-stepping implementations in Definition 4.1 are all of this type. Typically in such situations there is an implementation layer $\mathcal{D}^\varepsilon$ on which the system may exist in more than one mode i, and its dynamics therefore depends not only on its state $\mathbf{x}$ but also on the current mode, which therefore appears in the update map $\Psi(\mathbf{x};i)$. In other implementations the dynamics in $\mathcal{D}^\varepsilon$ is instead governed by a transition rule that only depends on $\mathbf{x}$, i.e., the state lies in a transition 'mode' and not in any mode i.

Let $\Omega \subset \mathcal{D}^\varepsilon$ be a set of points inside the switching layer on which 'event$(\mathbf{x};\varepsilon)$ = true' is satisfied, and let $\mathbf{x}_1, \mathbf{x}_2, \ldots$ denote a set of points inside Ω visited by $\mathbf{x}(t)$ at times $t_1, t_2, \ldots$. Integrating between these provides a map

$$\mathbf{x}_n = \Phi(\mathbf{x}_{n-1};\varepsilon) \quad \text{where} \quad \Phi : \Omega \mapsto \Omega\,. \tag{6.11}$$

We have been deliberately vague in defining the set Ω and map Φ, as the form of both depends on the implementation. In Chap. 7 we will see how these definitions apply to the 'experiments' from Chap. 4. Still, with these definitions we can derive some useful results, in particular without knowing the map (6.11) explicitly, we can derive some implications for the dynamics in the switching layer.

For sliding to occur, the map Φ must have an invariant set on Ω, but that set need not be unique, and by implication the sliding dynamics need not be unique.

According to the map (6.11), $\mathbf{x}(t)$ evolves in increments along each field $\mathbf{f}^i$, e.g.,

$$\begin{aligned} \mathbf{x}_n &= \mathbf{x}_{n-1} + \mathbf{f}^i(\mathbf{x}_{n-1})\Delta t + O\left(\Delta t^2\right) \\ \text{if} \quad & \mathbf{x}_{n-1} \in \Omega_i(\mathbf{x}_{n-1}, \mathbf{x}_0, t_{n-1})\,. \end{aligned} \tag{6.12}$$

The mode i selected at each time increment can depend not only on the current state $\mathbf{x}_{n-1}$, but is also typically history dependent, depending on the initial state $\mathbf{x}_0$ and the current time t_{n-1}. The regions $\Omega_i \subset \Omega$ may therefore overlap. Because this selection has discontinuities wherever the mode i changes, this raises the possibility that attractors of the map can bifurcate in a far more arbitrary manner than in continuous or differentiable maps, able to make abrupt jumps in topology and periodicity (as in Fig. 2.2(ii), for example).

We can also derive the effect of an attractor on the system's dynamics. Let $\mathbf{x}(t)$ evolve along an ε-infinitesimal neighbourhood of the discontinuity thresholds for a time interval $[0,T]$, switching between modes $i \in Z_N = \{1,2,\ldots,m\}$, at a sequence of times $t_1, t_2, \ldots, t_r$, where $0 = t_0 < t_1 < t_2 < \cdots < t_r = T$. Thus $\mathbf{x}(t)$ evolves along a different vector field $\mathbf{f}^i$ in mode $i \in Z_N$ on each time interval $[t_{j-1}, t_j]$ for some $i = \{1,\ldots,m\}$ and $j \in \{1,\ldots,r\}$. Let μ_i denote the total proportion of the time T spent evolving along $\mathbf{f}^i$,

$$\mu_i = \frac{1}{T}\sum_{n=1}^{r}\begin{cases} t_n - t_{n-1} & \text{if } \mathbf{x}_n \in \mathcal{R}_n\,, \\ 0 & \text{if } \mathbf{x}_n \notin \mathcal{R}_n\,. \end{cases} \tag{6.13}$$

Let $\gamma_i(\mathbf{x}_i) = 1$ if $\mathbf{x}_n$ is currently in mode $\mathbf{f}^i$ and $\gamma_i(\mathbf{x}_i) = 0$ otherwise. Then the total change in $\mathbf{x}(t)$ over the time increment $T = \sum_{n=1}^{r}(t_n - t_{n-1})$ is

$$\begin{aligned} \frac{\Delta\mathbf{x}}{T} &= \frac{1}{T}\sum_{n=1}^{r}\Delta\mathbf{x}_{n-1} \\ &= \frac{1}{T}\sum_{n=1}^{r}\gamma_i(\mathbf{x}_{n-1})\mathbf{f}^i(\mathbf{x}_{n-1})(t_n - t_{n-1}) + O\left(\Delta t^2\right) \\ &= \sum_{i=1}\mu_i\mathbf{f}^i(\mathbf{x}) \; + \; O(T)\,, \end{aligned} \tag{6.14}$$

where $\mu_i \geq 0$ and $\sum_{i=1}^{N}\mu_i = 1$. In the limit $T \to 0$ this gives an effective equation of motion,

$$\dot{\mathbf{x}} = \mathbf{F}^{\rm co}(\mathbf{x};\mu_1,\ldots,\mu_N) := \sum_{i=1}\mu_i\mathbf{f}^i(\mathbf{x})\,. \tag{6.15}$$

Comparing to (5.5a), we see that as the coefficients μ_i vary over [0, 1] the effective vector field $\mathbf{F}^{\rm co}$ traces out the convex hull $\mathcal{F}(\mathbf{x})$ of the $\mathbf{f}^i$'s.

Using these effective equations of motion we can understand a basic separation of scales that distinguishes motion *across* and *along* the discontinuity threshold. Assume that $\mathbf{x}$ evolves along an ε-neighbourhood of the intersection of discontinuity thresholds $\sigma_1 = \cdots = \sigma_m = 0$. Define scaled coordinates $\mathbf{x} = (u, v, \ldots, w, \underline{x})$, where $u = \sigma_1/\varepsilon$, $v = y/\varepsilon, \ldots, w = \sigma_m/\varepsilon$, and $\mathbf{x} \in \mathbb{R}^{n-m}$. Let $\mathbf{F}^{\rm co} = (F^{\rm co}_u, \ldots, F^{\rm co}_w, \underline{F}^{\rm co})$, then $\dot{\mathbf{x}} = \mathbf{F}^{\rm co}$ becomes

$$\begin{aligned} \varepsilon\dot{u} &= F^{\rm co}_u(\varepsilon u,\ldots,\varepsilon w,\underline{x};\mu_1,\ldots,\mu_m) \\ &= F^{\rm co}_u(0,\ldots,0,\mathbf{x};\mu_1,\ldots,\mu_m) + O(\varepsilon)\,, \\ \vdots\;\; & \;\;\vdots \\ \varepsilon\dot{w} &= F^{\rm co}_v(\varepsilon u,\ldots,\varepsilon w,\underline{x};\mu_1,\ldots,\mu_m) \\ &= F^{\rm co}_w(0,\ldots,0,\mathbf{x};\mu_1,\ldots,\mu_m) + O(\varepsilon)\,, \end{aligned} \tag{6.16a}$$

$$\begin{aligned} \dot{\underline{x}} &= \underline{F}^{\rm co}(\varepsilon u,\ldots,\varepsilon w,\underline{x};\mu_1,\ldots,\mu_m) \\ &= \underline{F}^{\rm co}(0,\ldots,0,\mathbf{x};\mu_1,\ldots,\mu_m) + O(\varepsilon)\,. \end{aligned} \tag{6.16b}$$

(Here part (a) labels the fast equations $\varepsilon[\dot{\ }] = \ldots$ and (b) the slow equation $\dot{\underline{x}} = \ldots$). The $(u, \ldots, w)$ coordinates therefore evolve on a fast timescale $\tau = t/\varepsilon$. Denoting the derivative with respect to τ by a prime, the system instead becomes

$$\begin{aligned} u' &= F_u^{co}(0,\ldots,0,\mathbf{x};\mu_1,\ldots,\mu_m) + O(\varepsilon) \,, \\ &\vdots \quad \vdots \\ w' &= F_w^{co}(0,\ldots,0,\mathbf{x};\mu_1,\ldots,\mu_m) + O(\varepsilon) \,, \qquad (6.17a) \\ \underline{x}' &= O(\varepsilon) \,. \qquad (6.17b) \end{aligned}$$

When we simulate the system on the τ-timescale, the variables $(u,\ldots,w)$ evolve across the $O(\varepsilon)$ space of the switching layer, while the variables $\mathbf{x}$ remain quasi-static.

To identify the coefficients μ_i in (6.14), we therefore simulate (6.17) for a time interval T, keeping $\underline{x}$ fixed. If the time interval T can be taken long enough that the coefficients μ_i as calculated by (6.13) reach a steady state, their values define an effective equation of motion (6.15) at any given $\mathbf{x}$. If one of the μ_i takes a value of unity, with all $\mu_{j\neq i} = 0$, then the system is determined to have *crossed* the discontinuity threshold. If one or more μ_i settle to steady values between 0 and 1, then the simulation of the fast $(u,\ldots,w)$ subsystem of (6.17) must have reach a steady state or other attractor inside the switching layer, and is said to be *sliding* along the discontinuity.

The maps (6.11) and their invariants may not in general have any closed form expression that can be determined from the system (5.1), but must be discovered by simulation and approximated.

Although (6.16) is similar formally to (6.6), the latter describes a continuous flow on $\sigma_1 = \cdots = \sigma_m = 0$, while the former describes a hybrid implementation that jumps between the 2^m different modes specified by (6.16) when each μ_j takes a value 0 or 1 in the neighbourhood of $\sigma_1 = \cdots = \sigma_m = 0$.

6.3 Sliding Perspective III: Smoothed Implementations

If switching is implemented by a smooth process, then we can proceed by steps that are actually very similar to Sect. 6.1. This has the advantage that the analysis then follows standard methods of singular perturbation theory, but this familiarity disguises unobvious ambiguities that accompany smoothing. Though we can describe these to some extent here, we are still learning what kind of dynamics, and more specifically what kind of singularities, persists under smoothing.

We begin with the system $\dot{\mathbf{x}} = \mathbf{F}(\mathbf{x};\nu_1,\ldots,\nu_m)$ in terms of switching multipliers $\nu_j = \text{step}(\sigma_j)$. To smooth the discontinuity we simply make the replacement $\nu_j \mapsto \phi^{\varepsilon_j}(\sigma_j)$, where $\phi^{\varepsilon_j}(\sigma_j)$ is a smooth monotonic function, satisfying $\phi^{\varepsilon_j}(\sigma_j) = \text{step}(\sigma_j) + O\left(\varepsilon_j\right)$ for $|\sigma_j| \geq \varepsilon_j$.

Observe that for $|\sigma_j| \geq \varepsilon_j$ this definition gives $\phi^{\varepsilon_j}(\sigma_j) = \phi^1(\sigma_j/\varepsilon_j)$, so let assume this also holds for $|\sigma_j| < \varepsilon$. The dynamics of the quantity ν_j is found simply by differentiating and applying the chain rule, $\dot{\nu}_j = \varepsilon_j^{-1}(\partial\phi^{\varepsilon_j}/\partial\sigma_j)\dot{\sigma}_j = \hat{\varepsilon}_j^{-1}\dot{\mathbf{x}}\cdot\nabla\sigma_j = \hat{\varepsilon}_j^{-1}\mathbf{F}\cdot\nabla\sigma_j = \hat{\varepsilon}_j F_j$, where $\hat{\varepsilon}_j = \varepsilon_j(\partial\phi^{\varepsilon_j}/\partial\sigma_j)^{-1}$.

The result is formally that in (6.5)–(6.6), except that the definition of ν_j now differs and, more crucially, the small quantity $\hat{\varepsilon}_j$ is now a function of σ_j. We have

$$\hat{\varepsilon}_j(\sigma_j)\dot{\nu}_j = F_j(0,\ldots,0,\underline{x};\nu_1\ldots,\nu_m) + O(\varepsilon_1,\ldots,\varepsilon_m) \;, \tag{6.18a}$$
$$\dot{\underline{x}} = \underline{F}(0,\ldots,0,\underline{x};\nu_1,\ldots,\nu_m) + O(\varepsilon_1,\ldots,\varepsilon_m) \;. \tag{6.18b}$$

By the definition of ϕ^{ε_j} the quantity $\hat{\varepsilon}_j$ is non-zero on the layer, where $|\sigma_j| < \varepsilon_j$. We can refine this if we limit the derivative of ϕ^{ε_j} away from zero, for example, choose ϕ^{ε_j} such that $\partial\phi^{\varepsilon_j}/\partial\sigma_j > K$ for $|\sigma_j| < 1 - \varepsilon_j^p$, for fixed $K > 0$ and $p \geq 1$ such that $\varepsilon_j/K \to 0$ as $\varepsilon_j \to 0$. Then $\hat{\varepsilon}_j$ behaves as a small quantity $\hat{\varepsilon}_j/K = O\big(\varepsilon_j\big)$ for $\sigma_j \in [-1 + \varepsilon_j^p, +1 - \varepsilon_j^p]$.

We see that this is analogous to the system (6.6) obtained by piecewise-smooth methods, and in fact they can be shown to be equivalent in the limit $\varepsilon_1, \varepsilon_2 \to 0$, see [81].

Equations of the form (6.18) can be found in singular perturbation studies of climate and gene regulation, e.g., [96, 103]. An equivalent form in common use when using Sotomayor–Teixeira regularization [136] is to define a parameter $u_j = \sigma_j/\varepsilon_j$, to obtain instead

$$\varepsilon_j\dot{u}_j = F_j(0,\ldots,0,\underline{x};\nu_1\ldots,\nu_m) + O(\varepsilon_1,\ldots,\varepsilon_m) \;, \tag{6.19a}$$
$$\dot{\underline{x}} = \underline{F}(0,\ldots,0,\underline{x};\nu_1,\ldots,\nu_m) + O(\varepsilon_1,\ldots,\varepsilon_m) \;. \tag{6.19b}$$

In either case (6.18) or (6.19), analysis proceeds using standard concepts from geometric singular perturbation theory, see, e.g., [49, 86]. If we assume all of the ε_js are of the same order, that is every ratio $\varepsilon_i/\varepsilon_j$ is non-vanishing as $\varepsilon_i, \varepsilon_j \to 0$ for $i, j = 1,\ldots,m$, then the analysis is closely analogous to that of the piecewise-smooth system in Sect. 6.1. The switching layer of the implementation is given as in (6.4) treating the ν_j as variables, or the same expression with ν_j replaced by u_j in the alternative variables. The slow-fast system has a critical manifold, corresponding precisely to the sliding manifold $\mathcal{M}$ in (6.7), where the fast ν_j subsystem (6.6a) vanishes. According to Fenichel's theory [49], wherever $\mathcal{M}$ is normally hyperbolic with respect to the fast ν_j subsystem, for $\varepsilon_j > 0$, there exists an invariant manifold $\mathcal{M}_{\varepsilon_j}$ in an ε_j-neighbourhood of $\mathcal{M}$. The dynamics on $\mathcal{M}$ is precisely the sliding dynamics (6.8), and moreover the dynamics on $\mathcal{M}^{\varepsilon_j}$ is topologically equivalent to (6.8).

If all of the ε_js are of different orders, then the dynamics in the switching layer will be more intricate, involving a separation onto more timescales, but still falls under standard methods of singular perturbation theory. The case $\varepsilon_j = \varepsilon_1^j$ for $j = 1,\ldots,m$, for instance, falls under Fenichel's analysis in [49]. The author is not aware of any studies to date applying such many timescale dynamics to piecewise-smooth problems.

The Sotomayor–Teixeira approach of replacing the switching multipliers ν_j by smooth (but non-analytic) functions $\phi^{\varepsilon_j}(\sigma_j)$ can be weakened so that the functions $\phi^{\varepsilon_j}(\sigma_j)$ are analytic. It is then impossible for these functions to be constant outside the switching layer, so they require defining to $\phi^{\varepsilon_j}(\sigma_j) = \text{step}(\sigma_j) + E$ where E is small, for example, $E = O\left(\varepsilon_j\right)$ or $O\left(\varepsilon_j/\sigma_j\right)$ or $O\left(\mathrm{e}^{-\sigma_j/\varepsilon_j}\right)$. This is often the case in applications. One example is in [96], where $\phi^{\varepsilon_j}(\sigma_j) = \frac{1}{2} + \frac{1}{2}\arctan(\sigma_j/\varepsilon) = \mathrm{step}(\sigma_j)+O\left(\varepsilon_j/\sigma_j\right)$. Another example is in [103, 117], where $\phi^{\varepsilon_j}(\sigma_j) = Z(\sigma_j+1) = \mathrm{step}(\sigma_j) + O(\sigma_j^{1/\varepsilon_j})$ in terms of the Hill function $Z(w) = 1/(1+w^{-1/\varepsilon_j})$ [70]. Both of these functions are analytic and are asymptotic to step(σ_j) for large argument.

Good examples of these methods applied to genetic models like that of Sect. 4.2 can be found in [41, 76, 102, 103, 117]. They tell a story similar to that obtained by piecewise-smooth analysis, but an appreciation of possible hidden terms would bring more insight into the robustness of these studies. Hill functions as a class are sometimes used without rigorous justification from the biology, and in such cases the possible differences between alternate sigmoid functions can be calculated as hidden terms, including representing the different between Hill functions of different stiffnesses (different powers $1/\varepsilon_j$).

Hidden terms survive when we smooth a discontinuity, and have the interpretation that they vanish (asymptotically at least) outside the discontinuity threshold. If we smooth by replacing $\nu_j \mapsto \phi^{\varepsilon_j}(\sigma_j) = \mathrm{step}(\sigma_j) + E(\varepsilon_j)$, then the term $\nu(\nu-1)$ which from (5.16) typically characterizes hidden terms, simplifies to

$$\nu(\nu-1) \quad\mapsto\quad 2\,\mathrm{sign}(\sigma_j)E(\varepsilon_j) + \mathrm{e}^2(\varepsilon_j)\;.$$

Hence the hidden term is of order $E(\varepsilon_j)$, vanishing (asymptotically) outside the discontinuity threshold with E.

Hidden terms therefore allow us to distinguish between different kinds or rates of switching according to different methods of smoothing. For instance, consider the one-dimensional system

$$\dot{x} = F(x;\nu_{(r)}) = a(x) + \nu b(x)\;, \tag{6.20}$$

defined in terms of a switching multiplier

$$\nu \;=\; \nu_{(r)} := \lim_{\varepsilon\to 0}\phi^{\varepsilon}_{(r)}(x)\;, \tag{6.21}$$

for different smooth functions $\phi^{\varepsilon}_{(1)}(x)$, $\phi^{\varepsilon}_{(2)}(x)$, ..., such that $\nu_{(r)} \to \mathrm{step}(x)$ for any r. Does it matter how we choose the function $\phi^{\varepsilon}_{(r)}$, or do we always obtain the same piecewise-smooth system (6.20) in the limit $\varepsilon \to 0$?

As examples consider the following sigmoid quantities

$$\phi^{\varepsilon}_{(0)}(x) = \tfrac{1}{2} + \frac{x/\varepsilon}{2\sqrt{1+(x/\varepsilon)^2}} \tag{6.22a}$$

$$\phi^{\varepsilon}_{(1)}(x) = \tfrac{1}{2} + \frac{x/\varepsilon}{2\sqrt{1+(x/\varepsilon)^2}} + \frac{A(x)}{2\left(1+(x/\varepsilon)^2\right)^k}\;, \tag{6.22b}$$

$$\phi^{\varepsilon}_{(2)}(x) = \tfrac{1}{2} + \tfrac{1}{\pi}\arctan(x/\varepsilon)\;, \tag{6.22c}$$

$$\phi^{\varepsilon}_{(3)}(x) = \tfrac{1}{2} + \tfrac{1}{2}\tanh(x/\varepsilon)\;, \tag{6.22d}$$

where $k > 0$ and $A(x)$ is a smooth function of x.

We will show the following.

Lemma 6.2. *We can write each system* $\dot{x} = F(x; \nu_{(r)})$ *from* (6.20) *as*

$$\dot{x} = F(x; \nu_{(r)}) = F(x; \text{step}(x)) + H^{\varepsilon}_{(r)}(x) , \tag{6.23}$$

an asymptotic expansion whose tail satisfies $H^{\varepsilon}_{(r)}(x) \to 0$ *as* $\varepsilon \to 0$ *for* $x \neq 0$. *This can be re-written in an* ε*-independent form as*

$$\dot{x} = F(x; \nu_{(r)}) = F(x; \nu_{(0)}) + H_{(r)}(x; \nu_{(0)}) , \tag{6.24}$$

with hidden terms satisfying $H_{(r)}(x; 0) = H_{(r)}(x; 1) = 0$, *but with* $H_{(r)}(x; \nu)$ *begin non-vanishing in the layer* $|x| < \varepsilon$ *for* $r = 1, 2, 3$.

Proof. The proof is directly by asymptotic expansion and straightforward calculations, so for brevity we place some of the details in Appendix C. The expansions of the sigmoid functions $\phi^{\varepsilon}_{(r)}$ for large argument all take the form $\phi^{\varepsilon}_{(r)}(x) = \text{step}(x) + O(\varepsilon/x)$ (the precise expressions are given in Appendix C, and in fact for $r = 3$ the error is $O\left(e^{-\varepsilon/x}\right)$). Substituting these into (6.20) gives (6.23) with the tail of the expansion given by $H^{\varepsilon}_{(r)}(x) = O(\varepsilon/x)$.

This means that the systems (6.20) with (6.21) and (6.22) are equivalent for $x \neq 0$. For $x = 0$, however, these asymptotic series diverge. To compare the different systems at $x = 0$, the second part of the theorem instead seeks a form of the expansions that is independent of ε, by expressing them all in terms of $\nu_{(0)}$.

To do this we first rearrange (6.22a) to find that

$$x/\varepsilon = (\phi^{\varepsilon}_{(0)} - \tfrac{1}{2})/\sqrt{\phi^{\varepsilon}_{(0)}(1 - \phi^{\varepsilon}_{(0)})} ,$$

followed by substituting this into the expressions in (6.22) and then into (6.20). To obtain (6.24) for $r = 1$ is then just a matter of algebra. To obtain (6.24) for $r = 2, 3$, it is better to substitute into the asymptotic expansions for each $\phi^{\varepsilon}_{(r)}$, thus expressing them in terms of $\phi^{\varepsilon}_{(0)}$. We then replace $\phi^{\varepsilon}_{(0)}$ with $\nu_{(0)}$. The algebra is set out in Appendix C, giving (6.24) for $r = 1, 2, 3$, with

$$H_{(1)}(x; \nu) = (4h)^k A(x) b(x) , \tag{6.25a}$$

$$H_{(2)}(x; \nu) = \left\{h\, C_1(2\nu - 1) + \sqrt{h} C_2(2\nu - 1)\right\} b(x) , \tag{6.25b}$$

$$H_{(3)}(x; \nu) = \left\{h\, C_1(2\nu - 1) + e^{-|2\nu-1|/\sqrt{h}}\, C_3(2\nu - 1)\right\} b(x) , \tag{6.25c}$$

$$\text{where} \quad h = \nu(1 - \nu) ,$$

in terms of functions $C_i(2\nu - 1)$ that are finite valued for all $\nu \in [0, 1]$, given in Appendix C. Clearly $H_{(r)}(x; 0) = H_{(r)}(x; 1) = 0$ in each case. These expressions are now independent of ε and therefore remain well-defined in terms of $\nu_{(0)}$ as $\varepsilon \to 0$.

□

Compare the hidden terms in (6.25) with the expression (5.12) in (5.3). Note how the term "$\nu(1 - \nu)$" appears throughout the hidden terms $H_{(r)}(x; \nu)$ in (6.25), but also demonstrates more general forms that hidden terms can take than we derived by polynomial expansion in (5.16).

This merely demonstrates that a piecewise-defined function corresponds not to one unique function of a switching multiplier ν or limiting smooth function $\phi^{\varepsilon}(x)$, but a whole class of such functions. Comparing to (5.3) we see that the difference between the alternate smoothings $\mathbf{F}(\mathbf{x};\nu_{(r)})$ lies in hidden terms.

Our interest, of course, concerns the dynamical implications of these hidden terms, left behind from the asymptotic approximations above. It should be quite clear that they can affect the system's dynamics. There are examples in [78, 79, 81, 83] of hidden terms deciding whether solutions slide along or cross through a discontinuity threshold. The anomalous sliding we described in Sects. 1.2 and 1.3 came from hidden terms, and more generally that can take all manner of non-trivial forms. An interesting example is given by taking (6.22b) and letting $A(x)$ be a matrix. The following example shows how this can destabilize an equilibrium under smoothing. Consider the piecewise-linear problem

$$\begin{pmatrix}\dot{x}\\ \dot{y}\\ \dot{z}\end{pmatrix}=\begin{pmatrix}1\\0\\0\end{pmatrix}+\nu\begin{pmatrix}-3\\ ay+z\\ az-y\end{pmatrix},\qquad \nu=\mathrm{step}(x)\,, \tag{6.26}$$

for $a<0$, which has sliding modes satisfying $\nu=1/3$, with an attracting focus equilibrium at $y=z=0$. If we smooth this system by replacing ν with $\phi^{\varepsilon}_{(0)}$, then we obtain a topologically equivalent system, with an attracting focus equilibrium on an invariant manifold, where $\phi^{\varepsilon}_{(0)}=1/3$. Consider instead smoothing by replacing ν with $\phi^{\varepsilon}_{(1)}$, and let

$$A=c\begin{pmatrix}-1/3&0&0\\0&a&-1\\0&1&a\end{pmatrix}, \tag{6.27}$$

for small $c>0$. As this is now a smooth system it succumbs to standard stability analysis. The system has an equilibrium at $(x,y,z)=(x_*,0,0)$, where $\phi_{(0)}(x_*/\varepsilon)=\frac{1}{3}+\frac{8c}{27}+O\left(c^2\right)$. This has eigenvalues $-\frac{3}{\varepsilon}(1-\frac{4c}{9})\phi'_{(0)}(x_*/\varepsilon)$ and $\frac{1}{3}(a\pm i)+\frac{8c}{27}(3+a+3a^2\pm i)$ to order c^2. This implies that for

$$\frac{-3a}{8[(a+1)^2-\frac{5}{3}a]}<c<\tfrac{9}{4}\,,$$

the equilibrium will de-stabilize in the (y,z) directions, becoming a saddle-focus as depicted in Fig. 6.1.

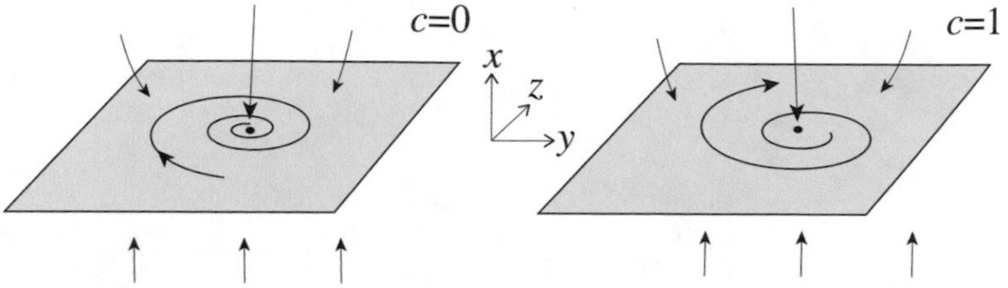

Fig. 6.1 A focus destabilized by smoothing

In this section we have seen some of the less obvious complexity of switching and sliding dynamics when considered from different viewpoints, expressed through layers, nonlinearity, and implementations. Let us now return to see what insight these give us into the ambiguities of the examples in Chap. 4.

Chapter 7
The Three Experiments Revisited

We now apply the framework of Chaps. 5 and 6 to make sense of the numerical experiments in Sects. 4.1 and 4.3.

7.1 Filippov's Paradox Revisited

First let us analyse the system (4.1) from the piecewise-smooth perspective. We could use the multiplier $\lambda = \text{sign}(x)$ from the original problem, but for consistency with the previous sections we will use $\nu = \frac{1}{2} + \frac{1}{2}\lambda = \text{step}(x)$. In terms of ν (4.1) becomes

$$\begin{pmatrix} \dot{x} \\ \dot{y} \end{pmatrix} = \begin{pmatrix} \frac{1}{2} - 2\nu \\ \frac{1}{3} + (2\nu - 1)^3 \end{pmatrix},$$

where $\nu = \text{step}(x)$ for $x \neq 0$ and $\nu \in [0, 1]$ for $x = 0$. Expressing this in the form of (5.14) from (5.3), we have

$$\begin{pmatrix} \dot{x} \\ \dot{y} \end{pmatrix} = \nu \begin{pmatrix} -\frac{3}{2} \\ \frac{4}{3} \end{pmatrix} + (1 - \nu) \begin{pmatrix} \frac{1}{2} \\ -\frac{2}{3} \end{pmatrix} + \mathbf{H}(\nu), \tag{7.1}$$

where $\mathbf{H}(\nu) = \nu(\nu - 1) \begin{pmatrix} 0 \\ 4(2\nu - 1) \end{pmatrix}$. We need only concern ourselves with the dynamics on $x = 0$.

Following Sect. 6.1, we first blow up the discontinuity threshold

$$\mathcal{D} = \left\{(x, y) \in \mathbb{R}^2 : x = 0\right\},$$

into a switching layer

$$\mathcal{D}^\varepsilon = \{(\nu, y) \in [0, 1] \times \mathbb{R}\},$$

M. R. Jeffrey, *Modeling with Nonsmooth Dynamics*,
Frontiers in Applied Dynamical Systems: Reviews and Tutorials 7,
https://doi.org/10.1007/978-3-030-35987-4_7

by defining $x = \varepsilon\nu \to 0$ as $\varepsilon \to 0^+$. Substituting $x = \varepsilon\nu$ into (7.1), on $x = 0$ for $\varepsilon \to 0^+$ this gives a two-timescale dynamical system

$$\begin{pmatrix} \varepsilon\dot{\nu} \\ \dot{y} \end{pmatrix} = \begin{pmatrix} \frac{1}{2} - 2\nu \\ -\frac{2}{3} + 2\nu \end{pmatrix} + \nu(\nu-1)\begin{pmatrix} 0 \\ 4(2\nu-1) \end{pmatrix}, \tag{7.2}$$

for $\nu \in (0, 1)$. Upon reaching $x = 0$, motion in (7.2) is dominated by the fast ν dynamics on the timescale t/ε, which is given, rescaling $t/\varepsilon = \tau$ and denoting the fast τ time derivative with a prime, by

$$\begin{pmatrix} \nu' \\ y' \end{pmatrix} = \begin{pmatrix} \frac{1}{2} - 2\nu \\ O(\varepsilon) \end{pmatrix}, \qquad \nu \in (0, 1)\,. \tag{7.3}$$

This tells us that ν contracts fast toward an invariant set $\nu = 1/4$ (where $\dot{\nu} = 0$), which according to Sect. 6.1 defines a sliding manifold $\mathcal{M} = \{(\nu, y) \in [0, 1] \times \mathbb{R} : \nu = 1/4\}$.

Finally, on $\mathcal{M}$ the dynamics is given by substituting $\nu = 1/4$ back into the vector field (7.2), with the result $(\dot{\nu}, \dot{y}) = (0, 5/24)$, giving the solution depicted in Fig. 7.1(i). (Substituting $\nu = 1/4$ directly into (7.1) gives, equivalently, $(\dot{x}, \dot{y}) = (0, 5/24)$).

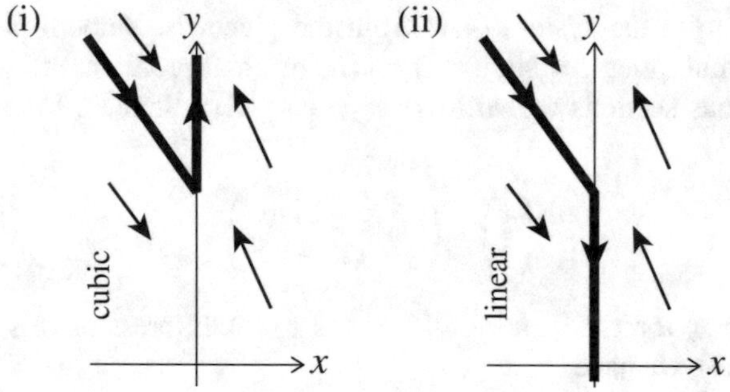

Fig. 7.1 A piecewise-constant vector field switching between two modes, showing the dynamics implied by piecewise-smooth analysis in (i) the nonlinear expression or (ii) the linear expression excluding hidden terms

If we neglect the hidden term, however, by setting $\mathbf{H} \equiv 0$ to obtain a linear switching (or 'Filippov') system, when we substitute $\nu = 1/4$ into (7.2) (or directly into (7.1)) we obtain $(\dot{\nu}, \dot{y}) = (0, -1/6)$ (or $(\dot{x}, \dot{y}) = (0, -1/6)$), giving the solution depicted in Fig. 7.1(ii).

Comparing these to Fig. 4.1, it appears that the smooth implementation follows the nonlinear dynamics of Fig. 7.1(i), while the time stepping, hysteretic, and delayed implementations follow the linear dynamics of Fig. 7.1(ii). A noisy implementation appears to be able to follow either the linear or nonlinear dynamics for small enough or large enough noise amplitude, respectively. Let us analyse these 'test implementations' (as defined in Definition 4.1) using the concepts from Chap. 6.

Following Sect. 6.3, to implement the switch by smoothing (7.1), we replace $\nu = \text{step}(x)$ with a smooth monotonic function $\phi(x/\varepsilon)$ for some small $\varepsilon > 0$, such that $\lim_{\varepsilon\to 0} \phi(x/\varepsilon) = \text{step}(x)$, giving

$$\begin{pmatrix} \dot{x} \\ \dot{y} \end{pmatrix} = \begin{pmatrix} \frac{1}{2} - 2\nu \\ -\frac{2}{3} + 2\phi(\frac{x}{\varepsilon}) \end{pmatrix} + \phi(\tfrac{x}{\varepsilon})(\phi(\tfrac{x}{\varepsilon}) - 1)\begin{pmatrix} 0 \\ 4(2\phi(\frac{x}{\varepsilon}) - 1) \end{pmatrix} . \tag{7.4}$$

For $|x| > \varepsilon$ this vector field is just the piecewise-constant field illustrated in Fig. 7.1. To analyse the neighbourhood $x \sim \varepsilon$ define a fast variable $u = x/\varepsilon$, giving a two timescale system

$$\begin{pmatrix} \varepsilon\dot{u} \\ \dot{y} \end{pmatrix} = \begin{pmatrix} \frac{1}{2} - 2\phi(u) \\ -\frac{2}{3} + 2\phi(u) \end{pmatrix} + \phi(u)(\phi(u) - 1)\begin{pmatrix} 0 \\ 4(2\phi(u) - 1) \end{pmatrix} \tag{7.5}$$

on the layer $\mathcal{D}^\varepsilon$ (the region $|x| < \varepsilon$). Scaling time by ε yields the fast system

$$\begin{pmatrix} u' \\ y' \end{pmatrix} = \begin{pmatrix} \frac{1}{2} - 2\phi(u) \\ O(\varepsilon) \end{pmatrix} , \qquad \nu \in (0, 1) . \tag{7.6}$$

This defines a critical manifold $\tilde{\mathcal{M}} = \left\{(u, y) \in \mathbb{R}^2 : \phi(u) = 1/4\right\}$, such that $\tilde{\mathcal{M}}$ and the dynamics $(\dot{u}, \dot{y}) = (0, 5/24)$ on it are topologically equivalent to the sliding manifold $\mathcal{M}$ and its sliding dynamics. We thus obtain the dynamics depicted in Fig. 7.1(i), consistent with the nonlinear piecewise-smooth system. The reader may easily verify that neglecting the hidden term (i.e., setting $\mathbf{H} \equiv 0$), and then smoothing the linear switching system, we instead obtain $(\dot{u}, \dot{v}) = (0, -1/6)$ consistent with Fig. 7.1(ii).

Now let us implement the switch by discretizing the system (7.1). For $x \neq 0$ the vector field is piecewise-constant and, having zero measure, a fixed time-step numerical method is unlikely to hit $x = 0$ instead evolving in increments of

$$(x_n, y_n) = (x_{n-1}, y_{n-1}) + \mathbf{v}\Delta t, \tag{7.7a}$$

where

$$\mathbf{v} = \begin{cases} (\frac{1}{2}, -\frac{2}{3}) & \text{if } x < 0 , \\ (-\frac{3}{2}, \frac{4}{3}) & \text{if } x > 0 , \end{cases} \tag{7.7b}$$

with $\Delta t = t_n - t_{n-1}$. These will step repeatedly back and forth across $x = 0$ over a layer $\mathcal{D}^\varepsilon$, where $\varepsilon = \Delta t$. Following Sect. 6.2, we can derive the effective vector field

$$(x_n - x_0, y_n - y_0)/\Delta t) = \mu(\tfrac{1}{2}, -\tfrac{2}{3}) + (1 - \mu)(-\tfrac{3}{2}, \tfrac{4}{3}) . \tag{7.8}$$

From this we can see that if the solution slides along $x = 0$ then, by solving $x_n - x_0 = 0$, we should find $\mu = \frac{1}{4}$. Simulations indeed yield $\mu \approx 1/4$, calculated from (6.13) as the proportion of time spent evolving in the $\nu = 1$ mode. Substituting $\mu \approx 1/4$

back into the y_n equation implies a speed of sliding motion of $\dot{y} = (y_n - y_0)/\Delta t = -\frac{1}{6}$, in agreement with the 'linear' sliding motion in Fig. 7.1(ii).

This is a rather crude numerical integration method, however, and a more precise simulation scheme might detect when a solution enters the layer $|x| < \varepsilon$, then switch to the layer system (7.2) on $\mathcal{D}^\varepsilon$. Provided the time step Δt is small enough, specifically $\Delta t < \varepsilon$, we solve

$$\begin{aligned} v_n &= \tfrac{1}{2} - \tfrac{\Delta t}{2\varepsilon} + \tfrac{1}{2}(1 - \tfrac{\Delta t}{\varepsilon})\lambda_{n-1} \\ &= \tfrac{1}{4} + (v_0 - \tfrac{1}{4})(1 - \tfrac{\Delta t}{\varepsilon})^n \ \xrightarrow{\varepsilon \to 0}\ \tfrac{1}{4} \end{aligned} \tag{7.9}$$

$$\begin{aligned} y_n &= y_{n-1} + (\tfrac{1}{3} + (2v_{n-1} - 1)^3)\Delta t \\ &= y_0 + \tfrac{n}{3}\Delta t + \Delta t \sum_{i=0}^{n-1} \Big(-\tfrac{1}{2} + (2v_0 - \tfrac{1}{2})(1 - \tfrac{\Delta t}{\varepsilon})^i\Big)^3 \\ &= y_0 + \tfrac{5n}{24}\Delta t + \varepsilon P \ \xrightarrow{\varepsilon \to 0}\ y_0 + \tfrac{5n}{24}\Delta t \end{aligned} \tag{7.10}$$

for $\Delta t < \varepsilon$, where

$$P = (2v_0 - \tfrac{1}{2})\left\{\frac{3}{4} - \frac{3}{2}\frac{2v_0 - \frac{1}{2}}{2 - \frac{\Delta t}{\varepsilon}} + \frac{2v_0 - \frac{1}{2}}{3 - \frac{3\Delta t}{\varepsilon} + \frac{\Delta t^2}{\varepsilon^2}}\right\}.$$

Thus $\dot{y} \approx (y - y_0)/n\Delta t = \frac{5}{24}$, and while we have not provided this simulation here, it is clear that it would agree with the 'nonlinear' sliding in Fig. 7.1(i).

For $\Delta t > \varepsilon$ the series in (7.10) does not converge, instead v_n diverges until it leaves the layer as $v_n \notin [0, 1]$, after which we can consider λ_n as switching between ± 1. If x_n stays in the neighbourhood of $x_n = 0$, the proportion of time μ spent in $x_n < 0$ versus the proportion of time spent in $x_n > 0$, must be such that over long times $x_n \approx 0$, so $\Delta x \approx v(-\frac{3}{2}) + (1 - v)\frac{1}{2}$, implying $v = \frac{1}{4}$ (as above), but then $\Delta y \approx v\frac{4}{3} - (1 - v)(\frac{2}{3}) = -\frac{1}{6}n\Delta t$, hence $\dot{y} \approx -\frac{1}{6}$, giving again the 'linear' sliding in Fig. 7.1(ii).

Thus a fine numerical simulation would find the nonlinear behaviour of Fig. 7.1(i), while a crude simulation would find the linear behaviour of Fig. 7.1(ii). We could smooth the system first, giving (7.2), and only *then* discretize, and then we obtain a similar result, namely that for a large time step $\Delta t \gg \varepsilon$ solutions 'chatter' across the switching layer $\mathcal{D}^\varepsilon$, which is now the ε-neighbourhood of $x = 0$, following the linear behaviour above. For a small time step $\Delta t \ll \varepsilon$ solutions are able to evolve into the layer and approximate the nonlinear behaviour.

Implementing the switch using hysteresis between the two modes in $x \gtrless 0$, where switching takes place at some $x = \pm\varepsilon$, results in a very simple result to the large time-stepping implementation above. The time taken to travel from $x = -\varepsilon$ to $x = +\varepsilon$ with $v = 0$ is $\delta t^- = 4$, and for the return with $v = 1$ is $\delta t^+ = 4$, giving maps $y_n = y_{n-1} - \frac{2}{3}\delta t^-$ and $y_n = y_{n-1} + \frac{4}{3}\delta t^+$. Concatenating these, the second return map to the surface $x = -\varepsilon$ (or any other surface $x = \textit{constant}$ in the implementation layer) is $y_n = y_{n-1} - \frac{8}{9}$, in a time step $\Delta t = \delta t^+ + \delta t^- = 16/3$. Hence the speed of motion along the y direction is $(y_n - y_{n-1})/\Delta t = -\frac{8}{9}/\frac{16}{3} = -1/6$, giving the dynamics in Fig. 4.1(iii). This fits with the linear piecewise-smooth system as shown in Fig. 7.1(ii). The rigorous proof that the approximation in fact takes the form $x = \mathcal{O}(\varepsilon)$, $\dot{y} = -\frac{1}{6} + \mathcal{O}(\varepsilon)$, is a lengthy exercise, a general proof of which is in [18].

The analysis considering delayed or stochastic implementations is rather longer and more complex, but similar in outcome. One may also combine different implementations in a single model (assigning a different constant ε to each), and investigate whether linear or nonlinear effects prevail depending on which implementation dominates. Analytic studies are in some cases able to derive precise asymptotic balances that define where any implementation dominates, but only small steps in this direction have been taken, e.g., in [18, 83]. Much work remains to be done in this direction, but these elements reveal how detailed an understanding of the switch implementation can be found, and how this generally fits either with the linear or nonlinear regimes of the piecewise-smooth theory.

In hindsight, it may look as if it should be a simple matter to distinguish between the two behaviours in Fig. 7.1, and the practical situations in which one or other is more appropriate. The literature in this area tells a different story. Filippov's intention with (4.1) was to demonstrate the dangers of considering non-convex switching, but restricting the inclusion in (5.2) to being convex essentially prevents us from resolving non-trivial behaviour by considering nonlinearity in the layer. Filippov seems not to have been averse to considering non-convex sets, but was clear to warn that analysing them would need an approach going beyond that which he set out in [51].

Having revealed the deeper modelling that is possible in one of the simplest and yet important behaviours of nonsmooth systems—the most basic sliding—we can turn to more complex scenarios where similar ambiguities have more novel and less obvious implications.

7.2 Genes Revisited

In the two-gene regulatory system from Sect. 4.2, applying the piecewise-smooth analysis outlined in Sect. 6.1 to the system (4.2), the dynamics in the switching layers is given by

- on $x_2 = \theta_2$,

$$\begin{pmatrix} \dot{x}_1 \\ \varepsilon_2 \dot{\nu}_2 \end{pmatrix} = \begin{pmatrix} \nu_1 + \nu_2 - 2\nu_1\nu_2 - \gamma_1 x_1 \\ 1 - \nu_1\nu_2 - \gamma_2\theta_2 \end{pmatrix}, \tag{7.11}$$

  with $\nu_1 = \text{step}(x_1 - \theta_1)$ and $\nu_2 \in (0, 1)$. On $x_1 < \theta_1$ where $\nu_1 = 0$ the fast system is constant (and positive), so no sliding occurs. On $x_1 > \theta_1$ the fast ν_2 subsystem has fixed points defining a sliding manifold

$$\mathcal{M} = \{(x_1, \nu_2) \in \mathbb{R} \times [0, 1] \ : \ \nu_2 = 1 - \gamma_2\theta_2\} \ .$$

On $\mathcal{M}$ the dynamics is $\dot{x}_2 = \gamma_2\theta_2 - \gamma_1 x_1$, suggesting the sliding flow on $x_2 - \theta_2 = 0 < x_1 - \theta_1$ has an equilibrium at $x_1 = \gamma_2\theta_2/\gamma_1$, easily shown to be an attractor.

- on $x_1 = \theta_1$,

$$\begin{pmatrix} \varepsilon_1 \dot{\nu}_1 \\ \dot{x}_2 \end{pmatrix} = \begin{pmatrix} \nu_1 + \nu_2 - 2\nu_1\nu_2 - \gamma_1\theta_1 \\ 1 - \nu_1\nu_2 - \gamma_2 x_2 \end{pmatrix}, \tag{7.12}$$

with $\nu_1 \in (0, 1)$ and $\nu_2 = \text{step}(x_2 - \theta_2)$. This layer will be of less interest, but the reader may show that sliding takes place on $x_1 - \theta_1 = 0 \gtrless x_2 - \theta_2$, and evolves towards $x_1 = \theta_1$, $x_2 = \theta_2$, if we assume $\gamma_1\theta_1 < \gamma_2\theta_2$.

- at their intersection $x_1 - \theta_1 = x_2 - \theta_2 = 0$,

$$\begin{pmatrix} \varepsilon_1 \dot{\nu}_1 \\ \varepsilon_2 \dot{\nu}_2 \end{pmatrix} = \begin{pmatrix} \nu_1 + \nu_2 - 2\nu_1\nu_2 - \gamma_1\theta_1 \\ 1 - \nu_1\nu_2 - \gamma_2\theta_2 \end{pmatrix}, \tag{7.13}$$

with $(\nu_1, \nu_2) \in (0, 1) \times (0, 1)$. This entire planar flow is fast, and possesses a sliding manifold $\mathcal{M} = \left\{(\nu_1, \nu_2) \in [0, 1]^2 \ : \ \nu_i = \nu_i^*\right\}$, where

$$\begin{Bmatrix} \nu_1^* \\ \nu_2^* \end{Bmatrix} = \tfrac{1}{2} + \tfrac{1}{2}(1 + \gamma_1\theta_1 - 2\gamma_2\theta_2)\begin{Bmatrix} 1 \\ 1 \end{Bmatrix} \pm \tfrac{1}{2}\sqrt{d}\begin{Bmatrix} -1 \\ 1 \end{Bmatrix},$$

where $d = (\frac{1}{2}\gamma_1\theta_1 - \gamma_2\theta_2)^2 + \gamma_1\theta_1 - \gamma_2\theta_2$. This only exists where $d > 0$ and $(\nu_1^*, \nu_2^*) \in (0, 1)^2$. The two branches of $\mathcal{M}$ are an attracting focus and a saddle, which as d passes from positive to negative, disappear in a saddle-node bifurcation.

The resulting phase portraits are sketched in Fig. 7.2.

The result is that if $d < 0$ then there is a unique global attractor, at $x_1 = (1 - \gamma_2\theta_2)/\gamma_1$, $x_2 = 0$, and solutions may slide along or cross through the various discontinuity thresholds before reaching it. If $d > 0$, then there is also an attractor

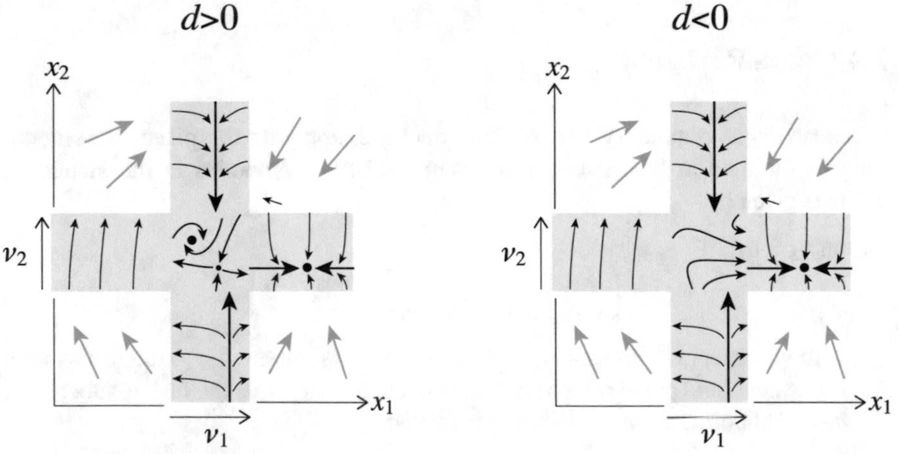

Fig. 7.2 Dynamics of a two-gene model, with switches at $x_1 = \theta_1$ and $x_2 = \theta_2$. These thresholds are blown up to reveal the dynamics in the layers $\nu_1 \in (0, 1)$ and $\nu_2 \in (0, 1)$. For $d > 0$ there exist a focus and a node in the layer at $x_1 = \theta_1$, $x_2 = \theta_2$, for $d < 0$ these disappear. These are sketched for parameter values $\theta_1 = \theta_2 = 1$, $\gamma_2 = 0.9$, and $\gamma_1 = 0.6$ (left), $\gamma_1 = 0.4$ (right)

at the intersection $x_1 = \theta_1$, $x_2 = \theta_2$, and many initial conditions, in particular any initial state with $x_1 < \theta_1$ (found by closer inspection of the phase portrait), will become stuck at $x_1 = \theta_1$, $x_2 = \theta_2$, in a sliding mode. This explains the results seen in Figs. 4.2 and 4.3, with sticking at the intersecting if $d > 0$, and evolution through it if $d < 0$.

The existence of sliding at $x_1 = \theta_1$, $x_2 = \theta_2$, in this example, but only for $d > 0$, means that whether or not solutions can cross through the intersection of the discontinuity thresholds depends on the system parameters. The change between $d > 0$ and $d < 0$ has no significant effect on the fields outside the thresholds, so the change in the dynamics this can only be seen by analysing the switching layer.

From the analysis in Sect. 7.1 we can expect that implementing the switch by smoothing gives dynamics that follows this piecewise-smooth behaviour on $x_1 - \theta_1 = 0 \neq x_2 - \theta_2$ and $x_2 - \theta_2 = 0 \neq x_1 - \theta_1$. Since there are no hidden terms here, there is only one outcome to consider. On the intersection $x_1 - \theta_1 = x_2 - \theta_2 = 0$ is a different story, however, as here the *bi*-linear term $v_1 v_2$ makes things non-trivial.

If we smooth the system following Sect. 6.3, by replacing each v_j with a smooth function $\phi_{(i)}^{\varepsilon_j}(x_j - \theta_j)$, then on the switching layer of the smoothing, given by

$$\mathcal{D}^{\varepsilon_1,\varepsilon_2} = \{(u_1, u_2) \in (-1, +1) \times (-1, +1)\} \ ,$$

we obtain in terms of a fast variable $u_i = (x_i - \theta_i)/\varepsilon_i$,

$$\begin{pmatrix} \varepsilon_1 \dot{u}_1 \\ \varepsilon_2 \dot{u}_2 \end{pmatrix} = \begin{pmatrix} \phi_1^{\varepsilon_1}(u_1) + \phi_2^{\varepsilon_2}(u_2) - 2\phi_1^{\varepsilon_1}(u_1)\phi_2^{\varepsilon_2}(u_2) - \gamma_1\theta_1 \\ 1 - \phi_1^{\varepsilon_1}(u_1)\phi_2^{\varepsilon_2}(u_2) - \gamma_2\theta_2 \end{pmatrix} . \tag{7.14}$$

We can readily see that this has equilibria closely corresponding to those of the piecewise-smooth system, at (u_1, u_2) such that $\phi_i^{\varepsilon_i}(u_i) = v_i^*$, existing only for $d > 0$. Thus the two behaviours seen with the smooth implementation in Figs. 4.2(i) and 4.3(i) are consistent with the two piecewise-smooth flows in Fig. 7.2.

The picture for hybrid implementations is rather more complicated. The convex hull (5.5a) says on $x_1 = \theta_1$, $x_2 = \theta_2$ only that

$$\begin{aligned} \begin{pmatrix} \dot{x}_1 \\ \dot{x}_2 \end{pmatrix} = \mu_{11}\begin{pmatrix} -\gamma_1\theta_1 \\ -\gamma_2\theta_2 \end{pmatrix} + \mu_{10}\begin{pmatrix} 1-\gamma_1\theta_1 \\ 1-\gamma_2\theta_2 \end{pmatrix} \\ + \mu_{01}\begin{pmatrix} 1-\gamma_1\theta_1 \\ 1-\gamma_2\theta_2 \end{pmatrix} + \mu_{00}\begin{pmatrix} -\gamma_1\theta_1 \\ 1-\gamma_2\theta_2 \end{pmatrix} , \end{aligned} \tag{7.15}$$

for some μ_{ij} such that $\mu_{11} + \mu_{01} + \mu_{10} + \mu_{00} = 1$ and $0 \leq \mu_{ij} \leq 1$. This contains families of vector fields that may carry solutions across the point $x_1 = \theta_1$, $x_2 = \theta_2$, and also contains the zero vector field

$$\begin{aligned} \begin{pmatrix} 0 \\ 0 \end{pmatrix} \in \mu_{11}\begin{pmatrix} -\gamma_1\theta_1 \\ -\gamma_2\theta_2 \end{pmatrix} + \mu_{10}\begin{pmatrix} 1-\gamma_1\theta_1 \\ 1-\gamma_2\theta_2 \end{pmatrix} \\ + \mu_{01}\begin{pmatrix} 1-\gamma_1\theta_1 \\ 1-\gamma_2\theta_2 \end{pmatrix} + \mu_{00}\begin{pmatrix} -\gamma_1\theta_1 \\ 1-\gamma_2\theta_2 \end{pmatrix} , \end{aligned}$$

which permits solutions to come to rest at $x_i = \theta_i$. Both are allowed by (7.15) and there is no way, based on the convex set, to select one possibility over any other.

The hybrid implementations cannot be solved analytically, but we can simulate them numerically and focus on their behaviour in the switching layers. Let us take the two examples of delay and hysteresis, the former of which showed behaviour depending on the parameters in Sect. 4.2, the latter seeming to 'round the corner' independent of parameters.

We implement a time delay at the discontinuity by defining the switching multipliers as $\nu_i = \text{step}\,(x_i(t-\varepsilon))$ for $\varepsilon = 0.2$, as in Figs. 4.2 and 4.3. The simulations are shown in Fig. 7.3 for $\gamma_1 = 0.4$ and 0.6, which correspond to $d = -0.01$ and 0.06. Consistent with the piecewise-smooth approach, for $d < 0$ the solution rounds the corner from the ε layer around x_1 to that around $x_2 = \theta_2$, eventually finding an attractor at $x_1 \approx \theta_1 + \theta_2\gamma_2/\gamma_1$, $x_2 \approx \theta_2$. For $d > 0$, instead the solution finds an attractor in the ε layer around $x_1 - \theta_1 = x_2 - \theta_2 = 0$, in this case a periodic cycle (whose period appears to be a multiple of 3 for these parameters). For other initial conditions solutions are still able to find a co-existing attractor at $x_1 \approx \theta_1 + \theta_2\gamma_2/\gamma_1$, $x_2 \approx \theta_2$.

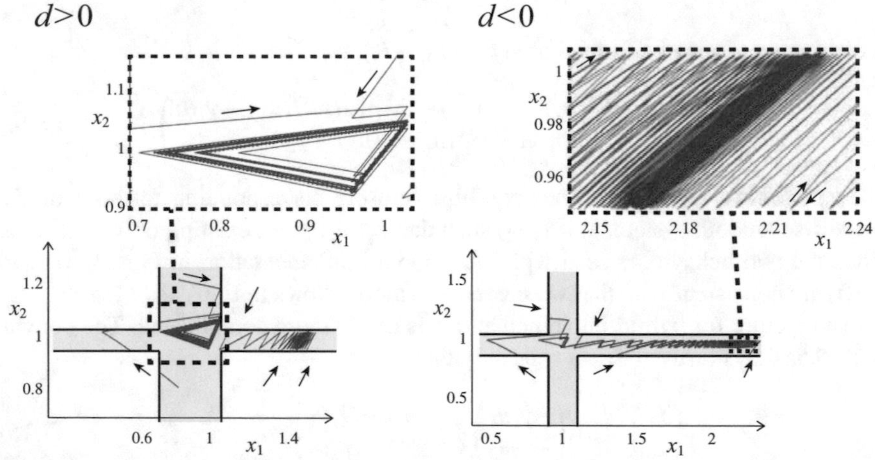

Fig. 7.3 A finer detail simulation of time delay implementations Figs. 4.2(vi) and 4.3(vi), with $\gamma_1 = 0.4$ (left) and $\gamma_1 = 0.6$ (right), showing solutions from several initial conditions. Approximate switching layers $\mathcal{D}^\varepsilon$ of the implementation are shaded. Left: solutions evolve either to an attractor at the intersection around $x_i \approx \theta_i$, or an attractor along $x_1 - \theta_1 > 0 \approx x_2 - \theta_2$. Right: all solutions evolve to an attractor along $x_1 - \theta_1 > 0 \approx x_2 - \theta_2$

We implement hysteresis at the discontinuity by switching ν_i as $0 \mapsto 1$ at $x_i = \theta_i + \varepsilon$, and $1 \mapsto 0$ at $x_i = \theta_i - \varepsilon$, with $\varepsilon = 0.2$ as in Figs. 4.2 and 4.3. The simulations are shown in Fig. 7.4 for $\gamma_1 = 0.4$ and 0.6, again corresponding to $d = -0.01$ and 0.06. The results of hysteresis are evidently not consistent with the piecewise-smooth approach, because regardless of the sign of d the solution rounds the corner from the ε layer around x_1 to that around $x_2 = 0$, eventually finding an attractor at

$x_1 \approx \gamma_2\theta_2/\gamma_1$, $x_2 \approx 0$. As stated in Sect. 4.2 there are isolated parameter values for which a hysteretic attractor forms in the ε layer around $x_1 = x_2 = 0$, but these are not typical.

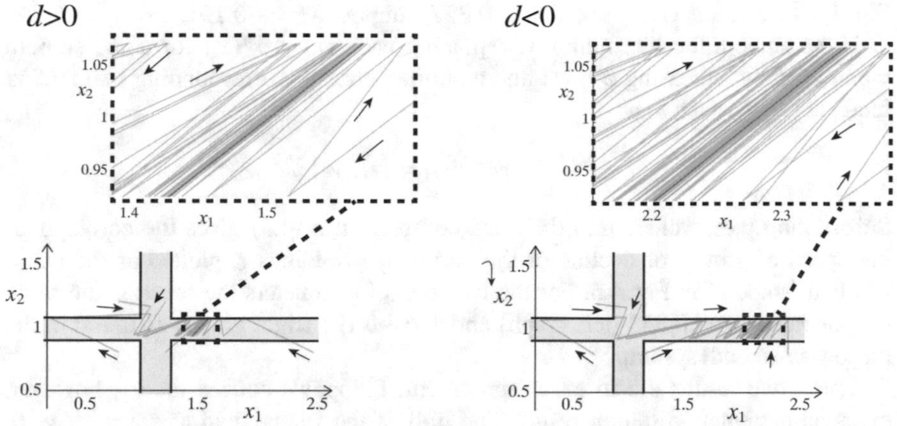

Fig. 7.4 A finer detail simulation of hysteretic delay implementations Figs. 4.2(iii) and 4.3(iii), with $\gamma_1 = 0.4$ (left) and $\gamma_1 = 0.6$ (right), showing solutions from several initial conditions, along with approximate switching layers $\mathcal{D}^\varepsilon$ shown shaded. In both cases all solutions evolve to an attractor along $x_1 > \theta_1$, $x_2 \approx \theta_2$

The exploration of other implementations is left to more focussed future work, hopefully driven by closer insight and experimental data concerning such non-idealities in the biological application. There is clearly much more precise modelling that can be achieved, though by perturbing around piecewise-smooth models it seems this can be done without overly complicating the models, or having unreasonable knowledge of the biophysical processes involved.

7.3 Investments Revisited

Lastly let us return to the game from Sect. 4.3. The system is such that solutions are attracted to the surfaces $x_1 = 0$ and $x_2 = 0$ (or their neighbourhood), and then to their intersection. So our only interest lies in finding the dynamics in the neighbourhood of the intersection $x_1 = x_2 = 0$.

The system (4.5) is in the form of (5.10). Treated as a piecewise-smooth system, we can use the methods of Sect. 6.1 to find that sliding modes exist in the layer $(\nu_1, \nu_2) \in (0,1) \times (0,1)$, and lie at $(\nu_1, \nu_2) = (\nu_1^*, \nu_2^*)$ such that

$$\{\nu_1^*, \nu_2^*\} = \left\{ \frac{\sqrt{d} - \gamma_1\gamma_2 + c(\rho_2 - \rho_1)}{2\gamma_1\rho_2}, \frac{\sqrt{d} - \gamma_1\gamma_2 - c(\rho_2 - \rho_1)}{2\gamma_2\rho_1} \right\}, \tag{7.16}$$

where $d = 4c\gamma_1\gamma_2\rho_2 + (\gamma_1\gamma_2 + c\rho_1 - c\rho_2)^2$. We then substitute these values into the third component in (4.5) to find the speed of growth or decline of the company's holdings z in the sliding mode. For the parameters in the left of Fig. 4.5 the result is $\left\{v_1^*, v_2^*\right\} = \{0.611, 0.475\}$, implying $\dot{z} = 0.033$, and for the parameters in the right of Fig. 4.5 the result is $\left\{v_1^*, v_2^*\right\} = \{0.402, 0.357\}$, implying $\dot{z} = -0.191$.

If we smooth the discontinuity, replacing each v_i by $\phi^{\varepsilon_i}(x_i)$ for some smooth monotonically increasing ϕ^{ε_i}, we find a similar attractor corresponding to a sliding mode at (v_1, v_2) such that

$$\{\phi^{\varepsilon_1}(v_1),\ \phi^{\varepsilon_2}(v_2)\} = \{v_1^*,\ v_2^*\}\ .$$

Substituting these values into the third component in (4.5) gives the corresponding speed of growth or decline of the company's holdings z, plotted as the curve labelled 'smooth' in Fig. 4.5. For the two sets of parameters the result is the same pair of values, $\dot{z} = 0.033$ (left graph) and $\dot{z} = -0.191$ (right graph) as found in the piecewise-smooth system.

These two results are in exact agreement. Filippov's convex theory, however, gives an extremely different result. The hull of the vector field at $x_1 = x_2 = 0$, applying (5.5a), is

$$\begin{pmatrix} \dot{x}_1 \\ \dot{x}_2 \\ \dot{z} \end{pmatrix} = \mu_{00}\mathbf{f}^{00} + \mu_{01}\mathbf{f}^{01} + \mu_{10}\mathbf{f}^{10} + \mu_{11}\mathbf{f}^{11}\ , \tag{7.17}$$

with a normalization condition $\mu_{00} + \mu_{01} + \mu_{10} + \mu_{11} = 1$. To obtain the convex hull of all possible vector fields, we let each μ_{ij} vary over its set $[0, 1]$. The boundaries of the hull correspond to vectors tangent to the intersection $x_1 = x_2 = 0$ when the vector fields in (7.17) are some $(\dot{x}_1, \dot{x}_2, \dot{z}) = (0, 0, F_3)$, with F_3 taking possible values (found by solving $\dot{x}_1 = \dot{x}_2 = 0$ in (7.17))

$$F_3^{ij} = f_3^{ij} + (f_3^{i+1,j} - f_3^{ij},\ f_3^{i,j+1} - f_3^{ij}) \cdot \begin{pmatrix} \mu_{i+1,j} \\ \mu_{i,j+1} \end{pmatrix} \tag{7.18}$$

$$\text{where} \qquad \begin{pmatrix} \mu_{i+1,j} \\ \mu_{i,j+1} \end{pmatrix} = -\begin{pmatrix} f_1^{i+1,j} - f_1^{ij} & f_1^{i,j+1} - f_1^{ij} \\ f_2^{i+1,j} - f_2^{ij} & f_2^{i,j+1} - f_2^{ij} \end{pmatrix}^{-1} \cdot \begin{pmatrix} f_1^{ij} \\ f_2^{ij} \end{pmatrix}$$

where $i, j \in \{0, 1\}$ and the indices $i + 1, j$, and $i, j + 1$ are taken modulo 2. This formula creates 8 possible values of F_3^{ij}, only two of which are valid limits of the hull, corresponding to the greatest and least values for which the coefficients μ_i satisfy $(\mu_{i+1,j}, \mu_{i,j+1}, 1 - \mu_{i+1,j} - \mu_{i,j+1}) \in [0, 1]^3$. The formula gives:

- $-0.330 \leq \dot{z} \leq 0.453$ for the parameter set that gives the left of Fig. 4.4, from $\{\mu_{00}, \mu_{11}\} = \{0.441, 0.435\} \Rightarrow \dot{z} = F_3^{10} = 0.453$ and $\{\mu_{01}, \mu_{10}\} = \{0.345, 0.490\}$ $\Rightarrow \dot{z} = F_3^{11} = -0.330$.
- $-0.543 \leq \dot{z} \leq 0.075$ for the parameter set that gives the right of Fig. 4.4, from $\{\mu_{10}, \mu_{01}\} = \{0.455, 0.357\} \Rightarrow \dot{z} = F_3^{00} = 0.075$ and $\{\mu_{11}, \mu_{00}\} = \{0.333, 0.643\}$ $\Rightarrow \dot{z} = F_3^{01} = -0.543$.

The different hybrid implementations should all follow dynamics within this hull as set out in Sect. 6.2. Before looking at the speed of motion along z let us look at the dynamics in the (x_1, x_2) plane.

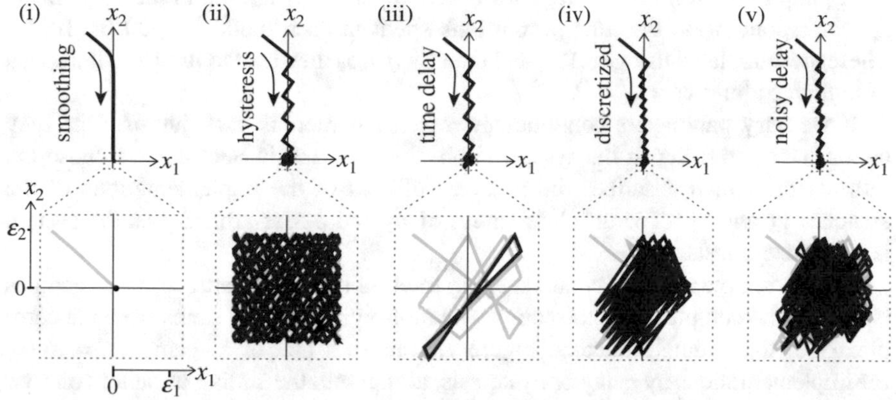

Fig. 7.5 A solution that evolves onto $x_1 \approx 0$ and then onto $x_1 \approx x_2 \approx 0$, whose neighbourhood is shown magnified, simulated with $\varepsilon_1 = \varepsilon_2$ for the parameter values in Fig. 4.5(left). [Fainter curves indicate transient of the solution for $0 \leq t \leq 25$, dark curves indicate $25 < t \leq 50$, indicating an attractor]

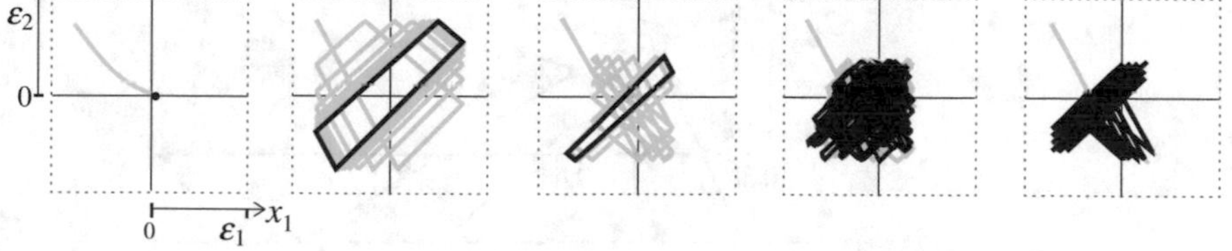

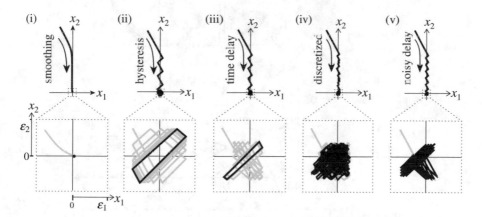

Fig. 7.6 As for Fig. 7.5, but simulated for the parameter values in Fig. 4.5(right)

In each simulation the flow finds an attractor in the neighbourhood of $x_1 = x_2 = 0$. If implemented by smoothing, then the attractor is a simple point (in the (x_1, x_2) plane), and otherwise, it is a more complex object whose extent in the x_1 and x_2 directions are of order ε_1 and ε_2, respectively. Hysteresis or time delay tends to lead to periodic or chaotic attractors.

As we change parameters within a given implementation, as we do from Figs. 7.5 and 7.6, the attractors undergo bifurcations. As remarked from the general form of such maps following (6.12), the fact they are only piecewise-differentiable permits these bifurcations to result in arbitrary changes in topology and period.

A jump in the form of the attractor is accompanied by a jump in the coefficients μ_i, corresponding to the time proportions spent in each mode $i = 00, 01, 10, 11$. These are calculated using (6.13), and then by (6.15) they lead to jumps in the speed of motion $\dot{z}$ along $x_1 \approx x_2 \approx 0$.

If we vary parameters continuously, we see numerous such jumps. One may choose any parameter in the system to observe this, but in Sect. 4.3 we chose the ratio $\varepsilon_2/\varepsilon_1$, which describes the relative stiffness of the implementations of the switches ν_1 and ν_2. (The absolute values of ε_1 and ε_2 are arbitrary as the system is piecewise-constant).

By varying the ratio $\varepsilon_2/\varepsilon_1$ and keeping track of the sliding attractors, it becomes clear how the complexity in the company's motion $\dot{z}$ in Fig. 4.5 stems from the complexity of the sliding attractors. Figure 7.7 shows a plot of $\dot{z}$ against $\omega = \varepsilon_2/\varepsilon_1$ for implementations by delay or hysteresis, along with the sliding attractors on four adjacent branches. The graphs in Fig. 4.5 show curves that consisting of smooth segments between sharp kinks, and each smooth segment is associated with a qualitatively different attractor. The examples shown demonstrate continuous changes of topology and/or periodicity, as well as sudden jumps to very high period.

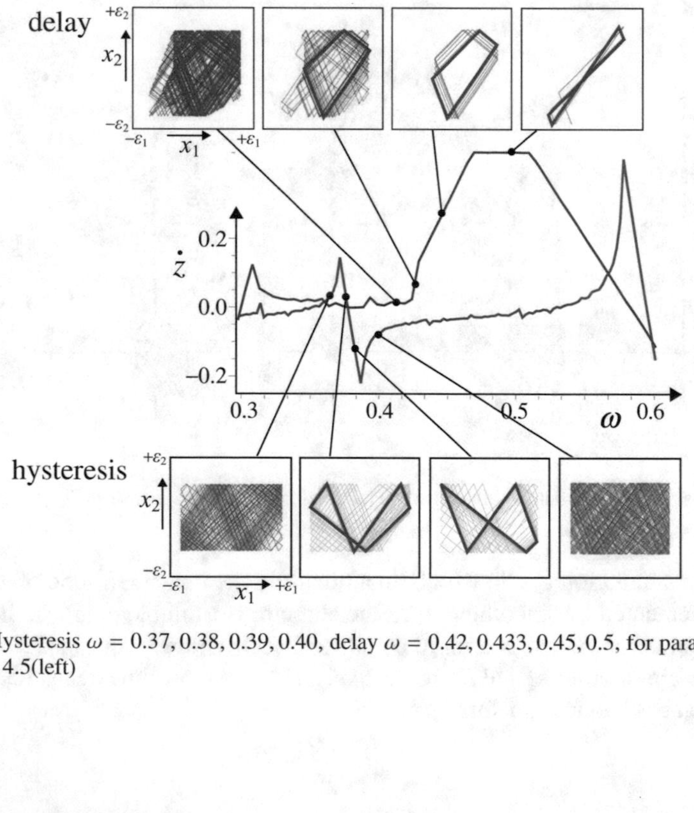

Fig. 7.7 Hysteresis $\omega = 0.37, 0.38, 0.39, 0.40$, delay $\omega = 0.42, 0.433, 0.45, 0.5$, for parameter values in Fig. 4.5(left)

Figure 7.7 is just a close-up on the region $0.3 < \omega < 0.6$ of the larger plot in Fig. 4.5, over which we see the full extent of how erratically $\dot{z}$ varies with ω. The surprise of Fig. 4.5 is that the various implementations lie not just inside the hull, but explore its entirety, so erratic is the variation in $\dot{z}$. The smooth and time-stepping implementations are not dependent on the ratio $\varepsilon_2/\varepsilon_1$ for this system (this happens because the vector field is piecewise-constant).

Ultimately the only way to understand the variations in $\dot{z}$ quantitatively is to study the maps (6.11), and being piecewise-constant, defined over four (possibly overlapping) regions on which the map is continuous, they are neither simple to express in a closed form, nor solvable in general. They may, however, be reducible in dimension, as is the case with hysteresis or delay.

Since in hysteresis the switch is implemented only at the boundaries $x_i\varepsilon_i$ of the switching layer, a map can be derived on that boundary, and will therefore be only one dimensional. We can coordinatize the switching layer boundary $\mathcal{D}^{\varepsilon_1,\varepsilon_2} = \{(x_1, x_2) : |x_i| \leq \varepsilon_i\}$, by defining some $\zeta \in [0, 1)$ such that $\zeta = 0, \frac{1}{4}, \frac{1}{2}, \frac{3}{4}$, correspond to the corners of $\mathcal{D}^{\varepsilon_1,\varepsilon_2}$ (see inset images in Fig. 7.8. The dynamics then takes the form of a circle map $\xi_n = \xi_{n-1} + \Psi(\xi_{n-1})$ where $\Psi : \mathcal{D}^{\varepsilon_1,\varepsilon_2} \mapsto \mathcal{D}^{\varepsilon_1,\varepsilon_2}$. The map is piecewise linear, with discontinuities due to the pre-images or images of the map lying on the corners, namely at $\xi = 0, \frac{1}{4}, \frac{1}{2}, \frac{3}{4}$, and at ξ values for which $\Psi(\xi) = 0, \frac{1}{4}, \frac{1}{2}, \frac{3}{4}$. As shown in [6], the second return map Ψ^2 is a continuous, but is still an eight-branch piecewise linear circle map. There are in fact two such maps, which we can think of as the clockwise and anti-clockwise maps around the circle, corresponding to the same dynamics.

The maps are calculated in Fig. 7.8 for two attractors from Fig. 7.7 (the attractor are shown upper right of each map). With only a slight change in the map between (i) and (ii), in which a part of the attractor touches one of the vertices in the map (i.e., the solution touches a corner of the layer $\mathcal{D}^{\varepsilon_1,\varepsilon_2}$), the attractor jumps abruptly between high and low period. Note how the high period attractor nonetheless retains a ghost of the low period attractor.

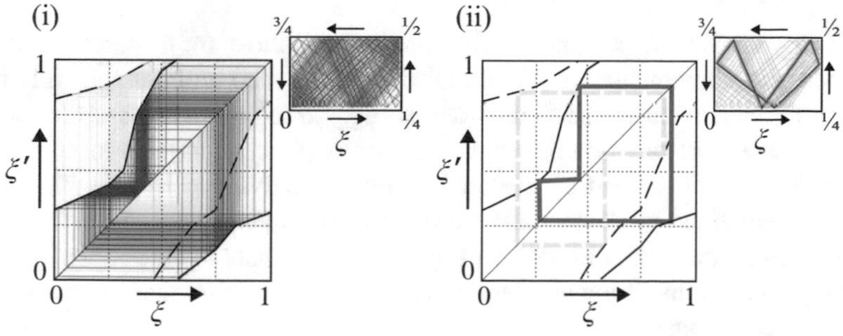

Fig. 7.8 The one-dimensional piecewise linear circle maps on the boundary of $\mathcal{D}^{\varepsilon_1,\varepsilon_2}$ for the hysteretic implementation of the investment game, calculated for two of the attractors from Fig. 7.7 with (i) $\omega = 0.37$, (ii) $\omega = 0.38$. The two (equivalent) clockwise and anti-clockwise maps are shown by full and dashed lines, along with their attractors shaded red and yellow, respectively. The corresponding attractor in the (x_1, x_2) plane is shown upper right of each map, including the ξ coordinatization around the boundary of $\mathcal{D}^{\varepsilon_1,\varepsilon_2}$

Bifurcations in piecewise linear maps like these are called *border collisions*. A great deal has been understood about border collision at a single discontinuity boundary, see, e.g., [56, 114, 130], and even how they respond to stochastic perturbation of the boundary [55], but for maps with many discontinuities there are no general patterns known concerning the bifurcations possible.

The map inside the layer is less simple for the case of delay, though it can in principle still be reduced to a one-dimensional map. The domain of that map, where switching occurs, is essentially the time-delayed image of the discontinuity thresholds $x_1 = 0$ and $x_2 = 0$. In Fig. 7.9 we plot out these discontinuity sets, by marking points in the plane where the vector field switches between the modes $\mathbf{f}^i$, for two of the attractors of the delay implementation from Fig. 7.7. Its shape is rather complicated by solutions that cross the surfaces $x_i = 0$ more than once during the delay time. The corresponding attractor is shown to the upper right of each plot. The high period attractor is seen to largely fill the enclosure of the map's domain. The bifurcations between attractors occur as the solution touches vertices of the one-dimensional domain.

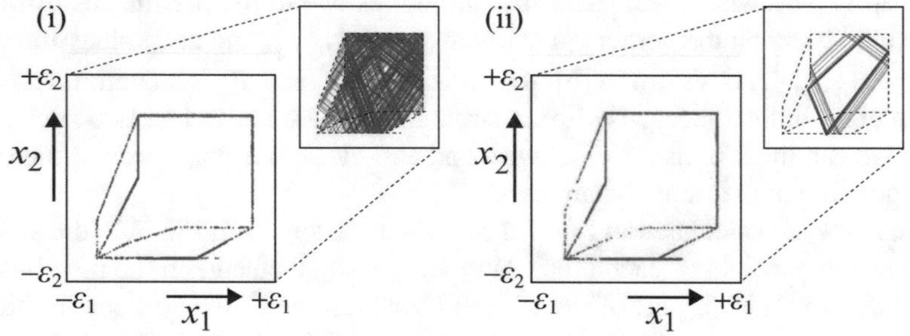

Fig. 7.9 The domain of the one-dimensional map of switching in the delay implementation of the investment game, calculated for two of the attractors from Fig. 7.7 with (i) $\omega = 0.42$, (ii) $\omega = 0.45$. The attractors are overlaid on these domains to the upper right of each plot

The only studies of such maps to date are perhaps those for hysteresis in [6, 84]. For stochastic implementations the dimension of the map will not generally be reducible in such a manner. The simulations in Fig. 4.6 suggest a tendency for noise to produce less erratic variation than hysteresis or delay, concentrated loosely around the attractor (7.16) derived by piecewise-smooth analysis or by smoothing. This was partly explained by numerically studying the probability density functions of a stochastic implementation in [84]. In Fig. 4.6 we simulate noise simply by adding random perturbations. If the problem is formulated more rigorously as a stochastic differential equation,

$$d\mathbf{x}(t) = \mathbf{F}(\mathbf{x}(t))dt + d\mathbf{W}(t, \varepsilon_1, \varepsilon_2), \tag{7.19}$$

where $\mathbf{W}(t, \varepsilon_1, \varepsilon_2)$ is a Brownian motion, the solutions form a smoother $\dot{z}$ curve than those in Fig. 4.6, and that curve lies close to the $\dot{z}$ graph of the piecewise-smooth system. Also as shown in [84], the probability density function then peaks at a value approximately consistent with the sliding attractor.

Even for the two player investment game, which takes the simplest form possible being only three dimensional and piecewise-constant, the resulting maps describing dynamics in the switching layer are one or two dimensional, with discontinuity thresholds that are themselves not simple (except in the case of hysteresis). They are therefore challenging to study exactly, having periodic behaviours corresponding to complex sliding attractors.

Nevertheless the quantitative behaviour of such systems is consistent, and proceeds in terms of sliding attractors inside switching layers, either of the blow-up or the implementation, and the overall result is clear. The dynamics at the intersection of two switches is unique as a piecewise-smooth system, or a smooth system, both evolving towards a sliding attractor (7.16). If switching is implemented by some hybrid process, the dynamics lies inside Filippov's convex hull (7.17), inside which it evolves to some sliding attractor. This attractor may be periodic or chaotic, and typically persists over only limited intervals of parameters, undergoing bifurcations in between that lead to abrupt changes in dynamics. The incompletely understood feature of these behaviours is how densely that dynamics will explore the convex set of all possibilities—Fig. 4.5 and other simulations, such as those in [6, 84], suggest a tendency to explore the entirety of the hull as parameters vary.

Chapter 8
Further Curiosities of Hidden Dynamics

Sections 4.3 and 7.3 introduced us to complex sliding modes as non-trivial attractors in the switching layer. The complexity arose due to non-idealities of the implementation, but it can also occur in ideal piecewise-smooth models due to hidden terms creating non-trivial sliding attractors. We give a brief overview here of a few examples presented in [82].

8.1 The Phenomenon of Jitter

In the simulations throughout Chap. 4, as the hybrid implementations attempt to follow sliding motion along the discontinuity thresholds, they exhibit slightly irregular motion which may be termed *chatter*. In the investment game we saw a much more violent behaviour associated with the non-ideality of implemented sliding, and we call this *jitter*. It arises if the attracting states in the switching layer responsible for sliding have multiple stable branches, or undergo bifurcations, that result in erratic variations or 'jitter' in the sliding dynamics.

Bifurcations in the sliding attractor (examples of which we saw in Figs. 7.5 and 7.6) cause dramatic changes in the times spent in each mode $\mathbf{f}^K$, such that the speed of travel along the discontinuity threshold changes in an erratic or 'jittery' manner.

The phenomenon also occurs when the attractor responsible for sliding has multiple branches or is chaotic, examples of which are given below for one or three discontinuity thresholds.

An illustration of jitter with one switch is given by the planar system

$$\begin{pmatrix} \dot{x} \\ \dot{y} \end{pmatrix} = \begin{pmatrix} y - 8\lambda + \sin(10\pi\lambda) \\ 1 + \frac{4}{5}\cos(4\pi\lambda) \end{pmatrix}, \tag{8.1}$$

with $\lambda = \text{sign}(x)$, as illustrated in Fig. 8.1(left), with the dynamics in the switching layer on $x = 0$ being

M. R. Jeffrey, *Modeling with Nonsmooth Dynamics*,
Frontiers in Applied Dynamical Systems: Reviews and Tutorials 7,
https://doi.org/10.1007/978-3-030-35987-4_8

$$\begin{pmatrix} \varepsilon\dot{\lambda} \\ \dot{y} \end{pmatrix} = \begin{pmatrix} y - 8\lambda + \sin(10\pi\lambda) \\ 1 + \frac{4}{5}\cos(4\pi\lambda) \end{pmatrix}, \tag{8.2}$$

for $\varepsilon \to 0$.

The invariant manifold on which sliding occurs is given by $y - 8\lambda + \sin(10\pi\lambda) = 0$, depicted in Fig. 8.1. It has turning points between attracting and repelling branches at every λ such that $\cos(10\pi\lambda) = 4/5\pi$. The sliding dynamics is given by

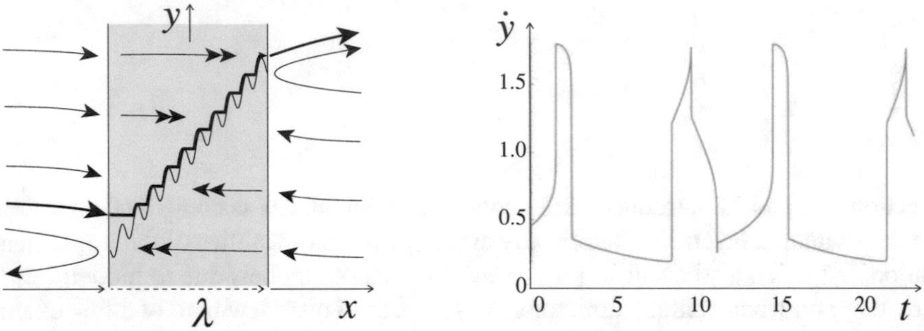

Fig. 8.1 Jitter in sliding motion, due to hidden jumps between branches of the sliding attractor. The piecewise-smooth flow is sketched along with the layer system (left). The jitter reveals itself in the vector field component $\dot{x}_2(t)$ (plotted right), which shows irregular jumps in sliding speed along the discontinuity threshold

$$\left.\begin{array}{l} 0 = y - 8\lambda + \sin(10\pi\lambda) \\ \dot{y} = 1 + \frac{4}{5}\cos(4\pi\lambda) \end{array}\right\} \quad : \quad \lambda \in (-1, +1) \ . \tag{8.3}$$

At each turning point a jump occurs to the next attracting branch of the sliding attractor, and because the $\dot{y}$ component depends on λ, this incurs an accompanying jump in $\dot{y}$ as simulated in Fig. 8.1(right). Thus the sliding solution evolves in bursts or ‘jitters’ along the discontinuity threshold, its vector field undergoing repeated jumps.

8.2 Hidden Oscillations and Chaos

If the attractors responsible for sliding can only be found by inspecting the dynamics inside the switching layer, and cannot be inferred solely from the modes $\mathbf{f}^K$ that exist outside the discontinuity thresholds, and we describe them as *hidden attractors*. The attractors involved can be much more complicated than those seen so far above, made of (possibly multiple-branch) invariant manifolds, as the following example shows.

Consider the one-switch system

$$\begin{pmatrix} \dot{x} \\ \dot{y} \\ \dot{z} \end{pmatrix} = \begin{pmatrix} y - cx \\ -\lambda^p - by + a\cos t \\ k(\lambda - z) \end{pmatrix}, \tag{8.4}$$

where $\lambda = \text{sign}(x)$, for constants a, b, c, some odd power p, and a large positive constant k, as proposed in [80]. The switching layer on $x = 0$ is

$$\begin{pmatrix} \varepsilon\dot{\lambda} \\ \dot{y} \\ \dot{z} \end{pmatrix} = \begin{pmatrix} y \\ -\lambda^p - by + a\cos t \\ k(\lambda - z) \end{pmatrix}. \tag{8.5}$$

The ideal piecewise-smooth system, with $\varepsilon \to 0$, has a stable periodic orbit satisfying $\lambda = a^{1/p}\cos^{1/p} t$, and z closely tracks this oscillating value of λ. For $\varepsilon > 0$ and $p > 1$ the system may exhibit chaos, for example, with $p = 3$ the $(\lambda, \dot{y})$ system is a Duffing oscillator [144, 145], and the chaotic attractor consists of oscillations of order ε around $\lambda \sim a^{1/p}\cos^{1/p} t$. This implies that chaos will only reveal itself under non-ideal implementations. Figure 8.2 plots $z(t)$ for a simulation of (8.4) for large k, implemented by smoothing for some small ε. A chaotic oscillation shrinks with ε towards the regular cycle with amplitude $a^{1/p}$ for $p = 3$. By contrast, for $p = 1$, there is only the simple periodic orbit with amplitude a.

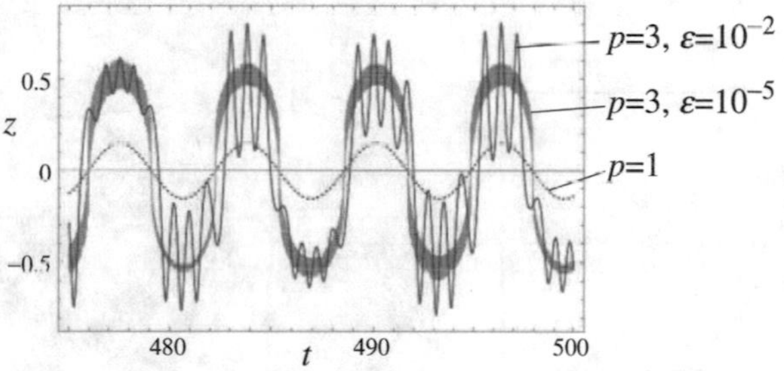

Fig. 8.2 Simulations of $z(t)$ from (8.4) for constants $p = 3$, $a = 0.15$, $b = 0.05$, $c = 0.1$, $k = 10^4$, and a smoothing implementation that replaces λ with $\lambda_\varepsilon = \tanh(x/\varepsilon)$, for $\varepsilon = 10^{-2}$ or $\varepsilon = 10^{-5}$. If we 'linearize' (8.4) by replacing λ^3 with λ we obtain the simple smaller oscillation (for any implementation)

Nonlinearities, therefore, manifest in two ways here, through the amplitude of the oscillation $\lambda \sim a^{1/p}\cos^{1/p} t$ originating in the hidden power of the switching multiplier $\lambda^p = \text{sign}(x)$ in (8.4), and through the chaos introduced by non-ideal implementation for $\varepsilon > 0$.

This is a contrived example that demonstrates nonlinearity by 'hiding' a Duffing oscillator inside the switching layer. An earlier example from [103], again motivated by genetic regulatory models, showed how a Lorenz attractor could be hidden inside the switching layer in a three-switch system. A slightly extended four-dimensional version of that model is

$$\begin{aligned}
\dot{y}_1 &= 10(\nu_2 - \nu_1) - 75y_1 , \\
\dot{y}_2 &= 2(2\nu_1 - 1)(7 - 15\nu_3) - \nu_2 + \tfrac{1}{2} - 75y_2 , \\
\dot{y}_3 &= 15(2\nu_1 - 2)(2\nu_2 - 1) - \tfrac{8}{3}\nu_3 - 75y_3 , \\
\dot{y}_4 &= \mu(\nu_1 - y_4) ,
\end{aligned} \tag{8.6}$$

where $\lambda_j = \text{sign}(x_j)$ for $j = 1, 2, 3$. This is based on the genetic models we saw already in Sects. 1.3 and 4.2, modelling protein concentrations $x_j = y_j + \frac{1}{4}$, released by genes that switch production on/off at discontinuity thresholds $y_i = 0$. The individual switching layer systems on $y_1 = 0$, $y_2 = 0$, $y_3 = 0$ are uninteresting. The origin $y_1 = y_2 = y_3 = 0$ is a global attractor, and there the switching layer dynamics is given by

$$\begin{aligned}
\varepsilon_1\dot{\lambda}_1 &= 10(\nu_2 - \nu_1) , \\
\varepsilon_2\dot{\lambda}_2 &= 2(2\nu_1 - 1)(7 - 15\nu_3) - \nu_2 + \tfrac{1}{2} , \\
\varepsilon_3\dot{\lambda}_3 &= 15(2\nu_1 - 2)(2\nu_2 - 1) - \tfrac{8}{3}\nu_3 , \\
\dot{y}_4 &= \mu(\nu_1 - y_4) .
\end{aligned} \tag{8.7}$$

If we let $\varepsilon_1 = \varepsilon_2 = \varepsilon_3$, this has a Lorenz attractor inside the layer $(\lambda_1, \lambda_2, \lambda_3) \in (0, 1) \times (0, 1) \times (0, 1)$. As shown in Fig. 8.3, while the protein concentrations y_i collapse to $y_i = 0$ (i), the multipliers ν_j enter into chaos (ii), and the coupled protein concentration y_4 following ν_1 likewise enters into chaos (iii).

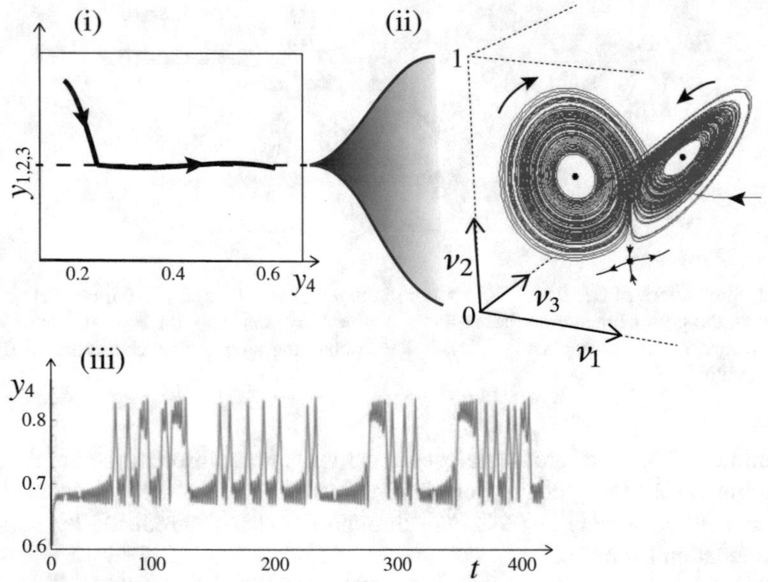

Fig. 8.3 Jitter in the system (8.6). As y_1, y_2, y_3, collapse to zero (i), the switching multipliers ν_j enter a Lorenz attractor (ii) in the switching layer around $y_1 = y_2 = y_3 = 0$, to which the variable y_4 is coupled (iii)

Such examples are intended only to hint at the range of behaviours that are possible in piecewise-smooth models, through stability changes, bifurcations, and chaos, affecting the attractors that constitute sliding, given nonlinear dependence on switching multipliers and their hidden effect inside the switching layer.

Chapter 9
Closing Remarks: Open Challenges

We have proposed here that piecewise-smooth dynamics can move forward, both in theory and application, by embracing a few less idealized notions, chiefly: *switching layers* as a blurring of ideal discontinuity thresholds, *sliding attractors* as a generalization of sliding modes, and *nonlinear* switching as a more general model of dynamics at a discontinuity. These elements allow us to distinguish between ideal piecewise-smooth analysis, and the results of implementing the discontinuity by a range of practically motivated processes. By bringing such non-idealities together as perturbations we may at last begin moving beyond Filippov's 'linear' theory and begin to fulfil his own larger vision.

Nonsmooth models are popular because they:

1. permit an intuitive geometric description of abrupt change;
2. reduce unsolvable problems to piecewise-solvable sub-problems;
3. give an idealized expression of switching.

Despite their simplicity and idealization, however, they:

1′. are sometimes numerically unstable in ways not fully understood;
2′. are often difficult and messy to analyze, case specific, and subject to a curse of dimensionality (see [58]);
3′. suffer from non-uniqueness.

In short, what nonsmooth systems gain in qualitative simplicity, they lose in analytic simplicity, generality, and determinacy. Nevertheless they have become ever more popular because of their wide applications and seeming ease of implementation. We have highlighted phenomena that reveal some unappreciated dangers of such models, while at the same time facilitating more detailed modelling and analysis.

The uncertainties of nonsmooth models present much greater modelling freedom than is currently taken advantage of. They are an admission of a lack of knowledge of the precise processes involved in a transition, leaving space in our equations to avoid over-modelling of aspects we cannot fix based on available data. They offer a ground-level model, the leading order of some more sophisticated approximation, to

M. R. Jeffrey, *Modeling with Nonsmooth Dynamics*,
Frontiers in Applied Dynamical Systems: Reviews and Tutorials 7,
https://doi.org/10.1007/978-3-030-35987-4_9

which we can add nonlinearities and implementations in a manner consistent with empirical data.

In this way, nonsmooth differential equations occupy a middle ground between differentiable equations on the one hand, and stochastic differential equations on the other hand, one fully deterministic and the other entirely non-deterministic. Instead, piecewise-smooth systems are *almost* deterministic, being differentiable in the neighbourhoods of almost all points in space, except at the discontinuity, where indeterminacy enters the picture, with many outcomes that can be studied via switching layers and hidden dynamics.

The peculiar outcome of the investment game says something fundamental about the mathematics of dynamic choice. The interaction of two players' decisions leads to volatile behaviour in the overall system, and this arises despite, and actually directly because, the investor's own dynamics fall onto an attractor. That volatility is breathed life by the incremental details of how each player's decisions are implemented.

The phenomenon of jitter reveals a surprising aspect of sliding dynamics, that despite providing a robust state of motion on a discontinuity threshold, it can nevertheless be responsible for very irregular behaviour. *Jitter* should not be confused with *chatter*, which is an entirely different phenomenon, its less technical usage meaning just the irregular jumping to-and-fro across the discontinuity threshold seen in Figs. 4.2, 4.3, and 7.1, and its more technical usage meaning a Zeno convergence of infinite impacts in finite time (see, e.g., [75]).

The definition of an implementation (5.3) of the system (5.1) is a simplification of Seidman's solution concept. In [124] Seidman loosens Filippov's definition of the convex set $\mathcal{F}(\mathbf{x})$ in a way that permits less idealized discontinuity thresholds, proving the existence of solutions when the set $\mathcal{F}^{\varepsilon}(\mathbf{x})$ contains all values that the vector field attains in an ε-neighbourhood of $\mathbf{x}$. This is a subtle change to Filippov's definition, but means that the discontinuity need only be defined as taking place in some ε-neighbourhood of a threshold, and $\dot{\mathbf{x}}$ need not have a unique, or indeed *any*, value at every point $\mathbf{x}$ in the neighbourhood, provided it has a value at *some* points in the neighbourhood. This is suggestive of a more general interpretation of nonsmooth dynamics as a modelling methodology, and suggests how to more formally define solutions through the *switching layers of implementation* introduced here. (The definition also has the advantage of extending to infinite dimensional systems, see [124]). The definition of switching layers bears some resemblance to boundary layers, to 'inflation' of differential equations (see [64]), and perhaps to other similar analytical methods or implementation techniques, any of which may bring welcome insights to aid in our further understanding of modelling with nonsmooth dynamics.

We summarize our main conclusion as follows. Given a piecewise-defined system (5.1), the inclusion (5.2) can be formed to prove the existence of solutions (using Filippov's theory [51]), but to specify or simulate them we need a more explicit formulation. Many applications are naturally described by (5.10) because they contain certain discontinuous parameters, such as currents being activated, or physical constants changing across material interfaces. Smoothing implementations as we discussed in Sect. 6.3 are exactly described by (5.10) and give dynamics similar to the piecewise-smooth system. Hybrid implementations of switching, of the types discussed in Sect. 6.2, turn out to be best described by the convex hull (5.5a), and if multiple switches are involved can exhibit great variation inside the hull, though noise tends to push the system towards the piecewise-smooth ideal. Switching layers provide a starting point to analyze such dynamics in detail.

Appendix A
Nonsmooth Models as Asymptotic Series

In Sect. 1.1 we sketched the derivation of a piecewise-smooth system, with leading order (1.2) and series approximation (1.11) for $\nu = \text{step}(x)$, from an asymptotic expansion with limiting behaviours (1.3) and uniform expression (1.11) in which $\nu \sim \text{step}(x)$. We flesh out the details of that derivation here. We then also show how to derive the asymptotics of the error function, as the method is worth illustrating, and derive the corresponding expression in the form (1.3).

A.1 The General Expression

First let us derive the expansion (1.3) as a general model for a piecewise-smooth system. Starting from (1.8), where ν is asymptotic to a step function, we can write (1.9) without loss of generality. Just as we noted that s might represent an expansion term x/ε or $e^{-\varepsilon/|x|}$, we may consider more specialized forms for ν provided they are asymptotic to the step function. These will change the precise expression of higher order terms, but not their general form.

Re-arranging (1.9) to write s in terms of ν for either $\pm x \gg \varepsilon$ separately, we have

$$s = \begin{cases} 1 - \nu + O\left((1-\nu)^2\right) & \text{for } x \gg +\varepsilon\,, \\ \nu + O\left(\nu^2\right) & \text{for } x \ll -\varepsilon\,. \end{cases} \tag{A.1}$$

Substituting these two limiting behaviours for s into the series (1.4) as appropriate gives for each term of the series

$$\begin{aligned} a_n(x;\varepsilon)s^n &= a_n(x;\varepsilon)\left(1-\nu+O\left((1-\nu)^2\right)\right)^n \\ &= \left(a_n(x;\varepsilon)+O\left((1-\nu)^2\right)\right)(1-\nu)^n \\ b_n(x;\varepsilon)s^n &= b_n(x;\varepsilon)\left(\nu+O\left(\nu^2\right)\right)^n \\ &= \left(b_n(x;\varepsilon)+O\left(\nu^2\right)\right)\nu^n\,, \end{aligned}$$

M. R. Jeffrey, *Modeling with Nonsmooth Dynamics*,
Frontiers in Applied Dynamical Systems: Reviews and Tutorials 7,
https://doi.org/10.1007/978-3-030-35987-4

yielding

$$A(x, s; \varepsilon) = \sum_{n=0}^{\infty} \tilde{a}_n(x; \varepsilon)(1-\nu)^n \quad \& \quad B(x, s; \varepsilon) = \sum_{n=0}^{\infty} \tilde{b}_n(x; \varepsilon)\nu^n , \tag{A.2}$$

where $\tilde{a}_0 = a_0$, $\tilde{b}_0 = b_0$, and for $n \geq 1$, $\tilde{a}_n = a_n + O(1-\nu)$, $\tilde{b}_n = b_n + O(\nu)$. Substituting these into (1.8) gives

$$\begin{aligned} F(x) &= \nu \sum_{n=0}^{\infty} \tilde{a}_n(x; \varepsilon)(1-\nu)^n + (1-\nu) \sum_{n=0}^{\infty} \tilde{b}_n(x; \varepsilon)\nu^n \\ &= \nu a_0(x; \varepsilon) + (1-\nu) b_0(x; \varepsilon) + \nu(1-\nu) G(x; \nu; \varepsilon) \end{aligned} \tag{A.3a}$$

where

$$G(x; \nu; \varepsilon) = \sum_{n=0}^{\infty} \left\{ \tilde{a}_{n+1}(x; \varepsilon)(1-\nu)^n + \tilde{b}_{n+1}(x; \varepsilon)\nu^n \right\} , \tag{A.3b}$$

where $\tilde{a}_n$ and $\tilde{b}_n$ are asymptotic to a_n and b_n. Letting $\varepsilon \to 0$ and defining $a_n(x) = a_n(x; 0)$, $b_n(x) = b_n(x; 0)$, we have finally

$$F(x) = \nu a_0(x) + (1-\nu) b_0(x) + \nu(1-\nu) G(x; \nu) \tag{A.4a}$$

where

$$G(x; \nu) = \sum_{n=0}^{\infty} \{ a_{n+1}(x)(1-\nu)^n + b_{n+1}(x)\nu^n \} . \tag{A.4b}$$

A.2 The Example of the Error Function

As an example we took F as given by the differential equation $\varepsilon^2 F'' = -xF'$ in (1.6). In this case F is known to be given by the Error function, with well-known asymptotics. For more complicated systems we would have to solve for F approximately, and a typical method would be roughly as follows.

We can integrate $\varepsilon^2 F'' = -xF'$ once quite simply, giving $\varepsilon^2 \log(F'/F'_0) = -\frac{1}{2}x^2$, where $F' = F'_0$ at $x = 0$. Exponentiating gives $F' = F'_0 \, e^{-x^2/2\varepsilon^2}$, and multiplying through by dx then gives $dF = F'_0 \, e^{-x^2/2\varepsilon^2} \, dx$. We can then integrate one more time,

$$\begin{aligned} \int_0^F dF &= F'_0 \int_0^x e^{-\xi^2/2\varepsilon^2} \, d\xi \\ &= F'_0 \left\{ \int_0^{\infty\,\mathrm{sign}(x)} - \int_x^{\infty\,\mathrm{sign}(x)} \right\} e^{-\xi^2/2\varepsilon^2} \, d\xi . \end{aligned} \tag{A.5a}$$

We have split the integral into two infinite pieces that are more readily calculated than the integral from 0 to x, with the 'sign' just ensuring that the integration contour passes through $x = 0$ only once. The left-hand side just evaluates to $F - F_0$ (where $F = F_0$ at $x = 0$). On the right-hand side the first integral evaluates simply as $\pm\varepsilon\sqrt{\pi/2}$ for $x \gtrless 0$, while the second must be tackled by a sequence of partial integrals.

The partial integration involves first treating the integrand $e^{-\xi^2/2\varepsilon^2}$ as the product $u.v' = (\frac{1}{\xi}).(\xi\, e^{-\xi^2/2\varepsilon^2})$, which equals $(uv)' - u'v = [-\frac{1}{\xi}\, e^{-\xi^2/2\varepsilon^2}]' - \frac{1}{\xi^2}\, e^{-\xi^2/2\varepsilon^2}$, giving an integrable first term and a remaining last term, which must again be tackled by partial integration. We proceed iteratively, on each successive integration taking the last term, of the form $\frac{1}{\xi^r}\, e^{-\xi^2/2\varepsilon^2}$, written as $u.v' = (\frac{1}{\xi^{r+1}}).(\xi\, e^{-\xi^2/2\varepsilon^2})$, which equals $(uv)' - u'v = [-\frac{1}{\xi^{r+1}}\, e^{-\xi^2/2\varepsilon^2}]' - \frac{r+2}{\xi^{r+2}}\, e^{-\xi^2/2\varepsilon^2}$, giving a series of integral terms and remainders

$$\begin{aligned}\frac{F-F_0}{\varepsilon F_0'} &= \sqrt{\tfrac{\pi}{2}}\,\mathrm{sign}(x) + \sum_{p=0}^{r}\Big[\frac{(-1)^p(2p-1)!!}{(\xi/\varepsilon)^{p+1}}\, e^{-\xi^2/2\varepsilon^2}\Big]_x^{\infty} \\ &\qquad - (2p-1)!!\int_x^{\infty}\frac{(-1)^r(r+2)}{\xi^{r+2}}\, e^{-\xi^2/2\varepsilon^2}\, d\xi \\ &= \sqrt{\tfrac{\pi}{2}}\,\mathrm{sign}(x) - e^{-x^2/2\varepsilon^2}\sum_{p=0}^{\infty}\frac{(-1)^p(2p-1)!!}{(x/\varepsilon)^{2p+1}}\,, \end{aligned} \tag{A.5b}$$

obtained by performed an infinite sequence of such partial integrations.

With $F_0 = c$ and $\varepsilon F_0' = \sqrt{\pi/2}$ from the example (1.6), we may write this as

$$\begin{aligned} F &= c + \mathrm{sign}(x) - \sqrt{\tfrac{2}{\pi}}\, e^{-x^2/2\varepsilon^2}\sum_{p=0}^{\infty}\frac{(-1)^p(2p-1)!!}{(x/\varepsilon)^{2p+1}} \\ &= \nu A(x,s;\varepsilon) + (1-\nu)B(x,s;\varepsilon)\,, \end{aligned} \tag{A.6}$$

where

$$A(x,s;\varepsilon) = c + 1 - \sqrt{\tfrac{2}{\pi}}\, e^{-x^2/2\varepsilon^2}\sum_{p=0}^{\infty}\frac{(-1)^p(2p-1)!!}{(x/\varepsilon)^{2p+1}}\,, \tag{A.7}$$

$$B(x,s;\varepsilon) = c - 1 - \sqrt{\tfrac{2}{\pi}}\, e^{-x^2/2\varepsilon^2}\sum_{p=0}^{\infty}\frac{(-1)^p(2p-1)!!}{(x/\varepsilon)^{2p+1}}\,, \tag{A.8}$$

with $\nu = \mathrm{step}(x)$.

The error function has particular importance as the universal mechanism of jumps in quantities in the asymptotic theory of Stokes discontinuities [17]. For more detailed theory and asymptotic methods behind the brief analysis summarized above for such integrals, see [39], and for a particularly readable account of these type of partial integration methods to obtain series expansions, see [68].

Appendix B
Simple Examples of Hulls, Canopies, and Indexing

It is a rather straightforward exercise to expand the hull or canopy expressions from Chap. 5 to make sense of their formulae for general number of regions N and number of switching thresholds m. To aid the reader let us expand the first two, using either of the indexing systems in (5.9).

For $N = 2$ and $m = 1$, the hull (5.5a) and canopy (5.7a) give expressions

$$\mathcal{F}(\mathbf{x}) = \mu_0\mathbf{f}^0(\mathbf{x}) + \mu_1\mathbf{f}^1(\mathbf{x}) ,$$
$$\mathcal{G}(\mathbf{x}) = (1 - \nu_1)\mathbf{f}^0(\mathbf{x}) + \nu_1\mathbf{f}^1(\mathbf{x}) ,$$

using the binary indexing from (5.9)(b) (which is trivial since $m = 1$), but these are identical since $\mu_0 = 1 - \mu_1$ by the normalization condition in (5.5a). (The numerical indexing from (5.9)(a) just increases the indices of μ_i and $\mathbf{f}^i$ here by 1). This does not mean that there is no distinction between linear and nonlinear switching systems for $m = 1$, however, as we could include hidden terms in the formula for $\mathcal{G}(\mathbf{x})$ proportional to $\nu_1(1 - \nu_1)$, as we introduce in Sect. 5.3.

For $N = 4$ and $m = 2$ these become

$$\mathcal{F}(\mathbf{x}) = \mu_1\mathbf{f}^1(\mathbf{x}) + \mu_2\mathbf{f}^2(\mathbf{x}) + \mu_3\mathbf{f}^3(\mathbf{x}) + \mu_4\mathbf{f}^4(\mathbf{x}) ,$$
$$\mathcal{G}(\mathbf{x}) = (1 - \nu_2)[(1 - \nu_1)\mathbf{f}^1(\mathbf{x}) + \nu_1\mathbf{f}^2(\mathbf{x})] + \nu_2[(1 - \nu_1)\mathbf{f}^3(\mathbf{x}) + \nu_1\mathbf{f}^4(\mathbf{x})] ,$$

in the numerical indexing (5.9)(a). These are no longer identical even though we can eliminate one μ_i, say $\mu_1 = 1 - \mu_2 - \mu_3 - \mu_4$, by the normalization condition. In the binary indexing (5.9)(b) these look like

$$\mathcal{F}(\mathbf{x}) = \mu_{00}\mathbf{f}^{00}(\mathbf{x}) + \mu_{10}\mathbf{f}^{10}(\mathbf{x}) + \mu_{01}\mathbf{f}^{01}(\mathbf{x}) + \mu_{11}\mathbf{f}^{11}(\mathbf{x}) ,$$
$$\mathcal{G}(\mathbf{x}) = (1 - \nu_2)[(1 - \nu_1)\mathbf{f}^{00}(\mathbf{x}) + \nu_1\mathbf{f}^{10}(\mathbf{x})] + \nu_2[(1 - \nu_1)\mathbf{f}^{01}(\mathbf{x}) + \nu_1\mathbf{f}^{11}(\mathbf{x})] ,$$

M. R. Jeffrey, *Modeling with Nonsmooth Dynamics*,
Frontiers in Applied Dynamical Systems: Reviews and Tutorials 7,
https://doi.org/10.1007/978-3-030-35987-4

with $\mu_{00} = 1 - \mu_{10} - \mu_{01} - \mu_{11}$. Now there is an immediate difference between the hull, which depends linearly on the μ_is, and the canopy which depends bi-linearly on the ν_js.

We could go on. What matters more, of course, is what we do with these formulae, and what dynamics they permit.

Appendix C
Deriving the Hidden Terms in Lemma 6.2

Taking the functions in (6.25) and expanding for large $|x|/\varepsilon$ gives

$$\frac{x/\varepsilon}{\sqrt{1+(x/\varepsilon)^2}} = \text{sign}(x)\{1 - \tfrac{1}{2}(\tfrac{\varepsilon}{x})^2 + \cdots + \tfrac{(-1)^n\Gamma(\frac{1}{2}+n)}{\sqrt{\pi}\Gamma(1+n)}(\tfrac{\varepsilon}{x})^{2n} + \ldots\}\,, \tag{C.1a}$$

$$\arctan(x/\varepsilon) = \tfrac{\pi}{2}\,\text{sign}(x) - \tfrac{\varepsilon}{x} + \tfrac{1}{3}(\tfrac{\varepsilon}{x})^3 + \cdots + \tfrac{(-1)^n}{2n+1}(\tfrac{\varepsilon}{x})^{2n+1} + \ldots\,, \tag{C.1b}$$

$$\tanh(x/\varepsilon) = \text{sign}(x)\{1 - 2\,\mathrm{e}^{-2|x|/\varepsilon} + \cdots + 2(-1)^n\,\mathrm{e}^{-2n|x|/\varepsilon} + \ldots\}\,. \tag{C.1c}$$

Introduce functions

$$c_1(x) = \frac{x/\varepsilon}{\sqrt{1+(x/\varepsilon)^2}} - \text{sign}(x)\,, \tag{C.2a}$$

$$c_2(x) = \tfrac{2}{\pi}\arctan(x/\varepsilon) - \text{sign}(x)\,, \tag{C.2b}$$

$$c_3(x) = \tanh(x/\varepsilon) - \text{sign}(x) \tag{C.2c}$$

such that

$$\tfrac{2}{\pi}\arctan(x/\varepsilon) - \frac{x/\varepsilon}{\sqrt{1+(x/\varepsilon)^2}} = c_2(x) - c_1(x)\,, \tag{C.3a}$$

$$\tanh(x/\varepsilon) - \frac{x/\varepsilon}{\sqrt{1+(x/\varepsilon)^2}} = c_3(x) - c_1(x)\,. \tag{C.3b}$$

Hence the functions $c_i(x)$ are themselves discontinuous at $x = 0$ but their differences are continuous functions.

We can use $\phi^{\varepsilon}_{(0)}$ from (6.22a) to substitute

$$x/\varepsilon = (\phi^{\varepsilon}_{(0)}(x) - \tfrac{1}{2})/\sqrt{\phi^{\varepsilon}_{(0)}(x)(1 - \phi^{\varepsilon}_{(0)}(x))}\,,$$

and hence express each c_i in terms of $\phi^{\varepsilon}_{(0)}$,

M. R. Jeffrey, *Modeling with Nonsmooth Dynamics*,
Frontiers in Applied Dynamical Systems: Reviews and Tutorials 7,
https://doi.org/10.1007/978-3-030-35987-4

$$c_1(x) = \operatorname{sign}(x)\left\{\tfrac{1}{2}(\tfrac{\varepsilon}{x})^2 - \cdots - \tfrac{(-1)^n\Gamma(\frac{1}{2}+n)}{\sqrt{\pi}\Gamma(1+n)}(\tfrac{\varepsilon}{x})^{2n} + \ldots\right\}$$

$$:= \phi^{\varepsilon}_{(0)}(x)(1-\phi^{\varepsilon}_{(0)}(x))C_1(1-2\phi^{\varepsilon}_{(0)}(x)) \tag{C.4a}$$

$$c_2(x) = \tfrac{2}{\pi}\left\{-\tfrac{\varepsilon}{x} + \tfrac{1}{3}(\tfrac{\varepsilon}{x})^3 + \cdots + \tfrac{(-1)^n}{2n+1}(\tfrac{\varepsilon}{x})^{2n+1} + \ldots\right\}$$

$$:= \sqrt{\phi^{\varepsilon}_{(0)}(x)(1-\phi^{\varepsilon}_{(0)}(x))}C_2(1-2\phi^{\varepsilon}_{(0)}(x)) \tag{C.4b}$$

$$c_3(x) = \operatorname{sign}(x)\left\{-2\,\mathrm{e}^{-2|x|/\varepsilon} + \cdots + 2(-1)^n\,\mathrm{e}^{-2n|x|/\varepsilon} + \ldots\right\}$$

$$:= \mathrm{e}^{-|1-2\phi^{\varepsilon}_{(0)}(x)|/\sqrt{\phi^{\varepsilon}_{(0)}(x)(1-\phi^{\varepsilon}_{(0)}(x))}}\,C_3(1-2\phi^{\varepsilon}_{(0)}(x)) \tag{C.4c}$$

in terms of functions

$$C_1(h) = \tfrac{2\operatorname{sign}(h)}{h^2}\left\{1 - \cdots - \tfrac{2(-1)^n\Gamma(\frac{1}{2}+n)}{\sqrt{\pi}\Gamma(1+n)}(\tfrac{1-h^2}{h^2})^{n-1} + \ldots\right\}, \tag{C.5a}$$

$$C_2(h) = \tfrac{4}{\pi h}\left\{-1 + \cdots + \tfrac{(-1)^n}{2n+1}(\tfrac{\sqrt{1-h^2}}{h})^{2n} + \ldots\right\}, \tag{C.5b}$$

$$C_3(h) = 2\operatorname{sign}(h)\left\{-1 + \cdots + (-1)^n\,\mathrm{e}^{-2(n-1)|h|/\sqrt{1-h^2}} + \ldots\right\}. \tag{C.5c}$$

The functions C_i inherit the boundedness of the functions c_i, along with the discontinuity which is now at $h = 0$.

References

1. M. Abramowitz, I. Stegun, *Handbook of Mathematical Functions* (Dover, New York, 1964)
2. V. Acary, H. de Jong, B. Brogliato, Numerical simulation of piecewise-linear models of gene regulatory networks using complementarity systems. Physica D **269**, 103–119 (2014)
3. M.A. Aizerman, F.R. Gantmakher, On the stability of equilibrium positions in discontinuous systems. Prikl. Mat. i Mekh. **24**, 283–293 (1960)
4. M.A. Aizerman, E.S. Pyatnitskii, Fundamentals of the theory of discontinuous systems I, II. Autom. Remote Control **35**, 1066–1079, 1242–1292 (1974)
5. J.C. Alexander, T.I. Seidman, Sliding modes in intersecting switching surfaces, I: blending. Houst. J. Math. **24**(3), 545–569 (1998)
6. J.C. Alexander, T.I. Seidman, Sliding modes in intersecting switching surfaces, II: hysteresis. Houst. J. Math. **25**(1), 185–211 (1999)
7. A.A. Andronov, A.A. Vitt, S.E. Khaikin, *Theory of Oscillations* (Fizmatgiz, Moscow, 1959, in Russian)
8. V. Avrutin, M. Schanz, On multi-parameteric bifurcations in a scalar piecewise-linear map. Nonlinearity **19**, 531–552 (2006)
9. J. Awrejcewicz, D. Sendkowski, Stick-slip chaos detection in coupled oscillators with friction. Int. J. Solids Struct. **42**, 5669–5682 (2005)
10. J. Awrejcewicz, L. Dzyubak, C. Grebogi, Estimation of chaotic and regular (stick-slip and slip-slip) oscillations exhibited by coupled oscillators with dry friction. Nonlinear Dyn. **42**, 383–394 (2005)
11. S. Banerjee, C. Grebogi, Border collision bifurcations in two-dimensional piecewise smooth maps. Phys. Rev. E **59**, 4052–4061 (1999)
12. F. Battelli, M. Feckan, Nonsmooth homoclinic orbits, Melnikov functions and chaos in discontinuous systems. Physica D **241**(22), 1962–1975 (2012)
13. A. Baule, E.G.D. Cohen, H. Touchette, A path integral approach to random motion with nonlinear friction. J. Phys. A **43**(2), 025003 (2010)

M. R. Jeffrey, *Modeling with Nonsmooth Dynamics*,
Frontiers in Applied Dynamical Systems: Reviews and Tutorials 7,
https://doi.org/10.1007/978-3-030-35987-4

14. C.M. Bender, S.A. Orszag, *Advanced Mathematical Methods for Scientists and Engineers I. Asymptotic Methods and Perturbation Theory* (Springer, New York, 1999)
15. M.T. Bengisu, A. Akay, Stick–slip oscillations: dynamics of friction and surface roughness. J. Acoust. Soc. Am. **105**(1), 194–205 (1999)
16. M.V. Berry, Stokes' phenomenon; smoothing a Victorian discontinuity. Publ. Math. Inst. Hautes Études Sci. **68**, 211–221 (1989)
17. M.V. Berry, Uniform asymptotic smoothing of Stokes's discontinuities. Proc. R. Soc. A **422**, 7–21 (1989)
18. C. Bonet, T.M. Seara, E. Fossas, M.R. Jeffrey, A unified approach to explain contrary effects of hysteresis and smoothing in nonsmooth systems. Commun. Nonlinear Sci. Numer. Simul. **50**, 142–168 (2017)
19. F.P. Bowden, D. Tabor, *The Friction and Lubrication of Solids* (Oxford University Press, Oxford, 1950)
20. Y. Braiman, F. Family, H.G.E. Hentschel, Nonlinear friction in the periodic stick-slip motion of coupled oscillators. Phys. Rev. B **55**(8), 5491 (1997)
21. C.J. Budd, A.R. Champneys, M. di Bernardo, Normal-form maps for grazing bifurcations in n-dimensional piecewise smooth dynamical systems. Physica D **160**, 222–254 (2001)
22. R. Burridge, L. Knopoff, Model and theoretical seismicity. Bull. Seism. Soc. Am. **57**, 341–371 (1967)
23. V. Carmona, F. Fernández-Sánchez, D.N. Novaes, A new simple proof for the Lum-Chua's conjecture (2019). arXiv:1911.01372
24. R. Casey, H. de Jong, J.L. Gouze, Piecewise-linear models of genetic regulatory networks: equilibria and their stability. J. Math. Biol. **52**, 27–56 (2006)
25. C.V. Chianca, J.S. Sá Martins, P.M.C. de Oliveira, Mapping the train model for earthquakes onto the stochastic sandpile model. Eur. Phys. J. B **68**, 549–555 (2009)
26. M. Cieplak, E.D. Smith, M.O. Robbins, Molecular origins of friction: the force on adsorbed layers. Science **265**(5176), 1209– 1212 (1994)
27. I. Clancy, D. Corcoran, State-variable friction for the Burridge-Knopoff model. Phys. Rev. E **80**, 016113 (2009)
28. E.A. Coddington, N. Levinson, *Theory of Ordinary Differential Equations* (McGraw-Hill, New York, 1955)
29. A. Colombo, M. Jeffrey, J.T. Lazaro, J.M. Olm (eds.), *Nonsmooth Dynamics, Extended Abstracts Spring 2016*. Trends in Mathematics (Springer, Berlin, 2017)
30. A.R. Crowther, R. Singh, Identification and quantification of stick-slip induced brake groan events using experimental and analytical investigations. Noise Control Eng. J. **56**(4), 235–255 (2008)
31. P.R. da Silva, I.S. Meza-Sarmiento, D.N. Novaes, Nonlinear sliding of discontinuous vector fields and singular perturbation (2017). arXiv:1706.07391
32. P.R. Dahl, A solid friction model, in *TOR-158(3107-18)* (The Aerospace Corporation, El Segundo, 1968)

33. E. Davidson, M. Levin, Gene regulatory networks (special feature). PNAS **102**(14), 4925 (2005)
34. M. di Bernardo, P. Kowalczyk, A. Nordmark, Bifurcations of dynamical systems with sliding: derivation of normal-form mappings. Physica D **170**, 175–205 (2002)
35. M. di Bernardo, C.J. Budd, A.R. Champneys, P. Kowalczyk, *Piecewise-Smooth Dynamical Systems: Theory and Applications* (Springer, Berlin, 2008)
36. L. Dieci, Sliding motion on the intersection of two surfaces: spirally attractive case. Commun. Nonlinear Sci. Numer. Simul. **26**, 65–74 (2015)
37. L. Dieci, L. Lopez, Sliding motion on discontinuity surfaces of high codimension. A construction for selecting a Filippov vector field. Numer. Math. **117**, 779–811 (2011)
38. L. Dieci, C. Elia, L. Lopez, A Filippov sliding vector field on an attracting co-dimension 2 discontinuity surface, and a limited loss-of-attractivity analysis. J. Differ. Equ. **254**, 1800–1832 (2013)
39. R.B. Dingle, *Asymptotic Expansions: Their Derivation and Interpretation* (Academic, London, 1973)
40. F. Dumortier, R. Roussarie, *Canard Cycles and Center Manifolds*, vol. 557 (American Mathematical Society, Providence, 1996)
41. R. Edwards, A. Machina, G. McGregor, P. van den Driessche, A modelling framework for gene regulatory networks including transcription and translation. Bull. Math. Biol. **77**, 953–983 (2015)
42. I. Eisenman, Factors controlling the bifurcation structure of sea ice retreat. J. Geophys. Res. **117**, D01111 (2012)
43. C.P. Fall, E.S. Marland, J.M. Wagner, J.J. Tyson, *Computational Cell Biology* (Springer, New York, 2002)
44. M.I. Feigin, Doubling of the oscillation period with C-bifurcations in piecewise continuous systems. J. Appl. Math. Mech. **34**, 861–869 (1970)
45. M.I. Feigin, On the generation of sets of subharmonic modes in a piecewise continuous systems. J. Appl. Math. Mech. **38**, 810–818 (1974)
46. M.I. Feigin, On the structure of C-bifurcation boundaries of piecewise continuous systems. J. Appl. Math. Mech **42**, 820–829 (1978)
47. M.I. Feigin, *Forced Oscillations in Systems with Discontinuous Nonlinearities* (Nauka, Moscow, 1994, in Russian)
48. M.I. Feigin, The increasingly complex structure of the bifurcation tree of a piecewise-smooth system. J. Appl. Math. Mech **59**, 853–863 (1995)
49. N. Fenichel, Geometric singular perturbation theory. J. Differ. Equ. **31**, 53–98 (1979)
50. A.F. Filippov, *Differential Equations with Discontinuous Right-Hand Side*, vol. 2 (American Mathematical Society Translations, 1964), pp. 199–231
51. A.F. Filippov, *Differential Equations with Discontinuous Righthand Sides* (Kluwer, Dordrecht, 1988) (original in Russian 1985)
52. I. Flügge-Lotz, *Discontinuous Automatic Control* (Princeton University Press, Princeton, 1953)

53. L. Gardini, F. Tramantona, V. Avrutin, M. Schanz, Border-collision bifurcations in 1d piecewise-linear maps and Leonov's approach. IJBC **20**(10), 3085–3104 (2010)
54. P. Glendinning, Robust chaos revisited. Eur. Phys. J. Spec. Top. **226**(9), 1721–1734 (2007)
55. P. Glendinning, The border collision normal form with stochastic switching surface. SIADS **13**(1), 181–193 (2014)
56. P. Glendinning, Bifurcation from stable fixed point to n-dimensional attractor in the border collision normal form. Nonlinearity **28**(10), 3457–3464 (2015)
57. P. Glendinning, Bifurcation from stable fixed point to two-dimensional attractor in the border collision normal form. IMA J. Appl. Math. **81**(4), 699–710 (2016)
58. P. Glendinning, M.R. Jeffrey, Grazing-sliding bifurcations, the border collision normal form, and the curse of dimensionality for nonsmooth bifurcation theory. Nonlinearity **28**, 263–283 (2015)
59. P. Glendinning, M.R. Jeffrey, An introduction to piecewise smooth dynamics, in *Advanced Courses in Mathematics* (CRM Barcelona, Birkhauser, Basel, 2019)
60. P. Glendinning, P. Kowalczyk, A. Nordmark, Multiple attractors in grazing-sliding bifurcations in an explicit example of Filippov type. IMA J. Appl. Math. **81**(4), 711–722 (2016)
61. P. Glendinning, M.R. Jeffrey, S. Webber, Pausing in piecewise-smooth dynamic systems. Proc. R. Soc. A **475**, 20180574 (2019)
62. M.R.A. Gouveia, J. Llibre, D.N. Novaes, C. Pessoa, Piecewise smooth dynamical systems: persistence of periodic solutions and normal forms. J. Differ. Equ. **260**(7), 6108–6129 (2016)
63. A. Granados, L. Alseda, M. Krupa, The period adding and incrementing bifurcations: from rotation theory to applications. SIAM Rev. **59**(2), 225–292 (2017)
64. L. Grüne, P.E. Kloeden, Discretization, inflation and perturbation of attractors, in *Ergodic Theory; Analysis and Efficient Simulation of Dynamical Systems* (2001), pp. 399–416
65. J. Guckenheimer, Review of [35] by M. di Bernardo, C.J. Budd, A.R. Champneys, P. Kowalczyk. SIAM Rev. **50**(3), 606–609 (2008)
66. N. Guglielmi, E. Hairer, Classification of hidden dynamics in discontinuous dynamical systems. SIADS **14**(3), 1454–1477 (2015)
67. P. Hartman, *Ordinary Differential Equations* (Wiley, New York, 1964)
68. J. Heading, *An Introduction to Phase-Integral Methods* (Methuen, 1962)
69. D. Hilbert, Mathematical problems. Bull. Am. Math. Soc. **8**(10), 437–479 (1902)
70. A.V. Hill, The possible effects of the aggregation of the molecules of haemoglobin on its dissociation curves. Proc. Physiol. Soc. **40**, iv–vii (1910)
71. E.J. Hinch, *Perturbation Methods* (Cambridge University Press, Cambridge 1991)

72. N. Hinrichs, M. Oestreich, K. Popp, On the modelling of friction oscillators. J. Sound Vib. **216**(3), 435–459 (1998)
73. S.J. Hogan, L. Higham, T.C.L. Griffin, Dynamics of a piecewise linear map with a gap. Proc. R. Soc. A. **463**, 49–65 (2007)
74. S.J. Hogan, M.E. Homer, M.R. Jeffrey, R. Szalai, Piecewise smooth dynamical systems theory: the case of the missing boundary equilibrium bifurcations. J. Nonlinear Sci. **26**(5), 1161–1173 (2016)
75. C. Hös, A.R. Champneys, Grazing bifurcations and chatter in a pressure relief valve model. Physica D **241**(22), 2068–2076 (2012)
76. D. Hudson, R. Edwards, Dynamics of transcription-translation networks. Physica D **331**, 102–113 (2016)
77. J. Ing, E. Pavlovskaia, M. Wiercigroch, S. Banerjee, Bifurcation analysis of an impact oscillator with a one-sided elastic constraints near grazing. Physica D **239**, 312–321 (2010)
78. M.R. Jeffrey, Dynamics at a switching intersection: hierarchy, isonomy, and multiple-sliding. SIADS **13**(3), 1082–1105 (2014)
79. M.R. Jeffrey, Hidden dynamics in models of discontinuity and switching. Physica D **273–274**, 34–45 (2014)
80. M.R. Jeffrey, Smoothing tautologies, hidden dynamics, and sigmoid asymptotics in piecewise smooth ODEs. Chaos **23**, 103125 (2015)
81. M.R. Jeffrey, The ghosts of departed quantities in switches and transitions. SIAM Rev. **60**(1), 116–136 (2017)
82. M.R. Jeffrey, *Hidden Dynamics: The Mathematics of Switches, Decisions, and Other Discontinuous Behaviour* (Springer, Berlin, 2019)
83. M.R. Jeffrey, D.J.W. Simpson, Non-Filippov dynamics arising from the smoothing of nonsmooth systems, and its robustness to noise. Nonlinear Dyn. **76**(2), 1395–1410 (2014)
84. M.R. Jeffrey, G. Kafanas, D.J.W. Simpson, Jitter in dynamical systems with intersecting discontinuity surfaces. IJBC **28**(6), 1–22 (2018)
85. H. Jiang, A.S.E. Chong, Y. Ueda, M. Wiercigroch, Grazing-induced bifurcations in impact oscillators with elastic and rigid constraints. Int. J. Mech. Sci. **127**, 204–214 (2017)
86. C.K.R.T. Jones, *Geometric Singular Perturbation Theory*, vol. 1609. Lecture Notes in Mathematics (Springer, New York, 1995), pp. 44–120
87. M. Kapitaniak, H. Vaziri, J. Paez Chavez, N. Krishnan, M. Wiercigroch, Unveiling complexity of drill–string vibrations: experiments and modelling. Int. J. Mech. Sci. **101–102**, 324–337 (2015)
88. G. Karlebach, R. Shamir, Modelling and analysis of gene regulatory networks. Nat. Rev. Mol. Cell Biol. **9**, 770–780 (2008)
89. L.E. Kollar, G. Stepan, J. Turi, Dynamics of piecewise linear discontinuous maps. Int. J. Bif. Chaos **14**(7), 2341–2351 (2004)
90. J. Krim, Friction at macroscopic and microscopic length scales. Am. J. Phys. **70**, 890–897 (2002)
91. P. Kukucka, Melnikov method for discontinuous planar systems. Nonlinear Anal. **66**, 2698–2719 (2007)

92. V. Kulebakin, On theory of vibration controller for electric machines. Theor. Exp. Electon **4** (1932, in Russian)
93. Yu.A. Kuznetsov, S. Rinaldi, A. Gragnani, One-parameter bifurcations in planar Filippov systems. Int. J. Bifurcation Chaos **13**, 2157–2188 (2003)
94. R. Lande, A quantitative genetic theory of life history evolution. Ecology **63**, 607–615 (1982)
95. J. Larmor, *Sir George Gabriel Stokes: Memoir and Scientific Correspondence*, vol. 1 (Cambridge University Press, Cambridge, 1907)
96. J. Leifeld, Perturbation of a nonsmooth supercritical Hopf bifurcation. 1–12 (2016). arXiv:1601.07930
97. R.I. Leine, H. Nijmeijer, *Dynamics and Bifurcations of Non-smooth Mechanical Systems*, vol. 18. Lecture Notes in Applied and Computational Mathematics (Springer, Berlin, 2004)
98. J. Llibre, P.R. da Silva, M.A. Teixeira, Sliding vector fields for non-smooth dynamical systems having intersecting switching manifolds. Nonlinearity **28**(2), 493–507 (2015)
99. J. Llibre, D.N. Novaes, M.A. Teixeira, Maximum number of limit cycles for certain piecewise linear dynamical systems. Nonlinear Dyn. **82**(3), 1159–1175 (2015)
100. T. LoFaro, Period-adding bifurcations in a one parameter family of interval maps. Math. Comput. Model. **24**, 27–41 (1996)
101. A.I. Lur'e, V.N. Postnikov, On the theory of stability of control systems. Appl. Math. Mech. **8**(3) (1944, in Russian)
102. A. Machina, R. Edwards, P. van den Dreissche, Sensitive dependence on initial conditions in gene networks. Chaos **23**, 025101 (2013)
103. A. Machina, R. Edwards, P. van den Dreissche, Singular dynamics in gene network models. SIADS **12**(1), 95–125 (2013)
104. T. Mestl, E. Plahte, S.W. Omholt, A mathematical framework for describing and analysing gene regulatory networks. J. Theor. Biol. **176**, 291–300 (1995)
105. E.M. Navarro-López, R. Suárez, Modelling and analysis of stick-slip behaviour in a drillstring under dry friction. *Congreso Anual de la AMCA* (2004), pp. 330–335
106. E.M. Navarro-López, R. Suárez, Practical approach to modelling and controlling stick-slip oscillations in oilwell drillstrings, in *Proceedings of the 2004 IEEE International Conference on Control Applications* (Taipei, 2004), pp. 1454–1460
107. Yu.I. Neimark, The method of point-wise mappings in theory of nonlinear oscillations. Nauka (1972, in Russian)
108. Yu.I. Neimark, S.D. Kinyapin, On the equilibrium state on a surface of discontinuity. Izv. VUZ. Radiofizika **3**, 694–705 (1960)
109. G. Nikolsky, On automatic stability of a ship on a given course. Proc. Central Commun. Lab. **1**, 34–75 (1934, in Russian)
110. A.B. Nordmark, Universal limit mapping in grazing bifurcations. Phys. Rev. E. **55**, 266–270 (1997)

111. D.N. Novaes, M.R. Jeffrey, Regularization of hidden dynamics in piecewise smooth flow. J. Differ. Equ. **259**, 4615–4633 (2015)
112. D.N. Novaes, E. Ponce, A simple solution to the Braga-Mello conjecture. IJBC **25**(1), 1550009 (2015)
113. J. Nussbaum, A. Ruina, A two degree-of-freedom earthquake model with static/dynamic friction. Pure Appl. Geophys. **125**(4), 629–656 (1987)
114. H.E. Nusse, E. Ott, J.A. Yorke, Border-collision bifurcations: an explanation for observed bifurcation phenomena. Phys. Rev. E **49**(2), 1073–1076 (1994)
115. H. Olsson, K.J. Astrom, C.C. de Wit, M. Gafvert, P. Lischinsky, Friction models and friction compensation. Eur. J. Control **4**, 176–195 (1998)
116. B.N.J. Persson, *Sliding Friction: Physical Principles and Applications* (Springer, Berlin, 1998)
117. E. Plahte, S. Kjøglum, Analysis and generic properties of gene regulatory networks with graded response functions. Physica D **201**, 150–176 (2005)
118. F. Plestan, Y. Shtessel, V. Brégeault, A. Poznyak, New methodologies for adaptive sliding mode control. Int. J. Control **83**(9), 1907–1919 (2010)
119. K. Popp, P. Stelter, Stick-slip vibrations and chaos. Philos. Trans. R. Soc. A **332**, 89–105 (1990)
120. T. Putelat, J.R. Willis, J.H.P. Dawes, On the seismic cycle seen as a relaxation oscillation. Philos. Mag. **28–29**(1–11), 3219–3243 (2008)
121. T. Putelat, J.H.P. Dawes, J.R. Willis, On the microphysical foundations of rate-and-state friction. J. Mech. Phys. Solids **59**(5), 1062–1075 (2011)
122. R. Röttger, U. Rückert, J. Taubert, J. Baumbach, How little do we actually know? On the size of gene regulatory networks. IEEE/ACM Trans. Comput. Biol. Bioinform. **9**(5), 1293–1300 (2012)
123. T.I. Seidman, The residue of model reduction. Lect. Notes Comput. Sci. **1066**, 201–207 (1996)
124. T.I. Seidman, Existence of generalized solutions for ordinary differential equations in Banach spaces. Int. J. Evol. Equ. **1**, 107–119 (2005)
125. T.I. Seidman, Some aspects of modeling with discontinuities. Int. J. Evol. Equ. **3**(4), 419–434 (2007)
126. O.V. Sergienko, D.R. Macayeal, R.A. Bindschadler, Stick–slip behavior of ice streams: modeling investigations. Ann. Glaciol. **50**(52), 87–94 (2009)
127. J. Shi, J. Guldner, V.I. Utkin, *Sliding Mode Control in Electro-Mechanical Systems* (CRC Press, Boca Raton, 1999)
128. L. Shih-Che, C. Yon-Ping, Smooth sliding-mode control for spacecraft attitude tracking maneuvers. J. Guid. Control. Dyn. **18**(6), 1345–1349 (1995)
129. D.J.W. Simpson, On resolving singularities of piecewise-smooth discontinuous vector fields via small perturbations. Discrete Contin. Dyn. Syst. **34**(9), 3803–3830 (2014)
130. D.J.W. Simpson, Border-collision bifurcations in R^n. SIAM Rev. **58**(2), 177–226 (2016)
131. D.J.W. Simpson, R. Kuske, The positive occupation time of Brownian motion with two-valued drift and asymptotic dynamics of sliding motion with noise. Stochastics Dyn. **14**(4), 1450010 (2014)

132. D.J.W. Simpson, R. Kuske, Stochastically perturbed sliding motion in piecewise-smooth systems. Discrete Contin. Dyn. Syst. Ser. B **19**(9), 2889–2913 (2014)
133. D.J.W. Simpson, J.D. Meiss, Shrinking point bifurcations of resonance tongues for piecewise-smooth, continuous maps. Nonlinearity **22**(5), 1123–1144 (2009)
134. J.-J.E. Slotine, W. Li, *Applied Nonlinear Control* (Prentice Hall, Englewood Cliffs, 1991)
135. M. Sorensen, S. DeWeerth, G. Cymbalyuk, R.L. Calabrese, Using a hybrid neural system to reveal regulation of neuronal network activity by an intrinsic current. J. Neurosci. **24**(23), 5427–5438 (2004)
136. J. Sotomayor, M.A. Teixeira, Regularization of discontinuous vector fields, in *Proceedings of the International Conference on Differential Equations* (Lisboa, 1996), pp. 207–223
137. D. Tabor, Triobology - the last 25 years. a personal view. Tribol. Int. **28**(1), 7–10 (1995)
138. M.A. Teixeira, Structural stability of pairings of vector fields and functions. Bull. Braz. Math. Soc. **9**(2), 63–82 (1978)
139. M.A. Teixeira, On topological stability of divergent diagrams of folds. Math. Z. **180**, 361–371 (1982)
140. M.A. Teixeira, Generic singularities of 3D piecewise smooth dynamical systems, in *Advances in Mathematics and Applications* (2018), pp. 373–404
141. M.A. Teixeira, P.R. da Silva, Regularization and singular perturbation techniques for non-smooth systems. Physica D **241**(22), 1948–1955 (2012)
142. G.A. Tomlinson, A molecular theory of friction. Philos. Mag. **7**(7), 905–939 (1929)
143. Y. Tsypkin, *Theory of Relay Control Systems* (Gostechizdat, Moscow, 1955, in Russian)
144. Y. Ueda, Randomly transitional phenomena in the system governed by Duffing's equation. J. Stat. Phys. **20**, 181–196 (1979)
145. Y. Ueda, Explosion of strange attractors exhibited by Duffing's equation, in *Nonlinear Dynamics* (New York Academy of Science, New York, 1980), pp. 422–434
146. V.I. Utkin, Variable structure systems with sliding modes. IEEE Trans. Autom. Control **22**, 212–222 (1977)
147. V.I. Utkin, *Sliding modes and their application in variable structure systems*, volume (Translated from the Russian). MiR (1978)
148. V.I. Utkin, *Sliding Modes in Control and Optimization* (Springer, Berlin, 1992)
149. V.I. Utkin, Comments for the continuation method by A.F. Filippov for discontinuous systems, part I–II. *Trends in Mathematics: Research Perspectives*, vol. 8 (CRM Barcelona; Birkhauser, 2017), pp. 177–188
150. Various, Special issue on dynamics and bifurcations of nonsmooth systems. Physica D **241**(22), 1825–2082 (2012)

151. J. Wojewoda, S. Andrzej, M. Wiercigroch, T. Kapitaniak, Hysteretic effects of dry friction: modelling and experimental studies. Philos. Trans. R. Soc. A **366**, 747–765 (2008)
152. J. Woodhouse, T. Putelat, A. McKay, Are there reliable constitutive laws for dynamic friction? Philos. Trans. R. Soc. A **373**, 20140401 (2015)
153. H. Xu, M.D. Mirmirani, P.A. Ioannou, Adaptive sliding mode control design for a hypersonic flight vehicle. J. Guid. Control. Dyn. **27**(5), 829–38 (2004)
154. F.R. Zypman, J. John Ferrante, M. Jansen, K. Scanlon, P. Abel, Evidence of self-organized criticality in dry sliding friction. J. Phys. Condens. Matter **15**, L191–L196 (2003)

Index

M. R. Jeffrey, *Modeling with Nonsmooth Dynamics*,
Frontiers in Applied Dynamical Systems: Reviews and Tutorials 7,
https://doi.org/10.1007/978-3-030-35987-4